INTEGRATING TELECOMMUNICATIONS INTO EDUCATION

Nancy Roberts
George Blakeslee
Maureen Brown
Cecilia Lenk

Integrating Telecommunications Into Education

Nancy Roberts
George Blakeslee
Maureen Brown
Cecilia Lenk

PRENTICE HALL
Englewood Cliffs, New Jersey 07632

Library of Congress Cataloging-in-Publication Data

Integrating telecommunications into education / Nancy Roberts ... [et al.].
p. cm.
Includes bibliographical references.
ISBN 0-13-468547-4
1. Telecommunications in education. I. Roberts, Nancy
LB1044.84.I58 1990
371.3'078--dc20
89-28756
CIP

Editorial/production supervision: Evalyn Schoppet
Cover design: 20/20 Services, Inc.
Manufacturing buyer: Peter Havens

Printed in the United States of America
10 9 8 7 6 5 4 3 2 1

ISBN 0-13-468547-4

Prentice-Hall International (UK) Limited, *London*
Prentice-Hall of Australia Pty. Limited, *Sydney*
Prentice-Hall Canada, Inc., *Toronto*
Prentice-Hall Hispanoamericana, S. A., *Mexico*
Prentice-Hall of India Private Limited, *New Delhi*
Prentice-Hall of Japan, Inc., *Tokyo*
Simon & Schuster Asia Ptd. Ltd., *Singapore*
Editora Prentice-Hall do Brazil, Ltda., *Rio de Janeiro*

To all the entrepreneurial risk-taking teachers whose experiences reported here are creating tomorrow's modes of learning and teaching.

Contents

Preface

Perhaps it's not quite a revolution, but technology is certainly moving quite rapidly into the schools. Computers, CD-ROM drives, videotape, and videodisc players are not going to end up in closets, at least not for long. Even television sets, which might have been closeted for a while, are being retrieved for use with VCRs.

The national reports published over the last fifteen years have portrayed a devastating picture of what is going on in our schools. These reports, and the press attention they have received, have greatly increased the general public's awareness of current problems with public school education. Moreover, people are slowly being convinced that the weakening of the United States economy, in part because of ill-prepared students, is already happening—it is not something that might happen at some future time if we don't act fast.

Technology has also been seeping, or in some cases flooding, into the work environment. Desktop computer screens that were used initially merely as a place to post handwritten notes are slowly being turned on and used for accessing data and sharing information. The truly global nature of information is now a reality. These events have caused people to think more carefully about contributions technology might make to the quality of education for our children.

Telecommunications is a key technology for globalizing information. Given our conviction in this, we saw a need to collect in one place some of the current telecommunications activities in schools, the how-to-do-it information needed, and a look at what technologies are on the horizon for those teachers who are going to change the seep of technology into the classroom into a flood. This book is by no means all one would want to know about telecommunications. It focuses primarily on computer-related telecommunications and not on other aspects of the field such as radio and television. We do hope, however, that it is more than enough to get you enthused and "up and running."

The book is divided into three parts to represent the three areas we believe are important to school personnel who want to integrate telecommunications into educational activities. Following the introduction in Chapter 1, the first part, "Getting Online," begins with Chapter 2, a tutorial to get the reader communicating with a free, educationally oriented bulletin board. Chapter 3 explains in detail how to introduce telecommunications to a class, specifically through telecommunications simulator software that requires no telephone connection. Local services are then described so that a teacher can begin the telecommunications experience with minimum costs. Chapter 4 explains the more high-powered and expensive commercial services available which are useful for a range of educational activities.

Part II of the book focuses on classroom activities that integrate telecommunications into the discipline areas. Chapter 5 shares with the reader a host of telecommunications activities that teachers around the country have used to enhance the teaching of language arts and foreign languages. Chapter 6 explores the use of telecommunications in the sciences and social studies. The many issues faced by school personnel in planning, budgeting, and managing the integration of telecommunications into a school system are discussed in Chapter 7.

More advanced topics are presented in the last part of the book. Chapter 8 guides the reader through the vocabulary that has developed in the field, in relation to both hardware and software. By explaining these terms, more information about telecommunications is presented. Chapter 9 gives the reader an overview of other sophisticated technologies that are being tried in educational settings. We believe these are the kinds of technological support that will be readily available in classrooms before the end of this century.

The book concludes with a Resource Section to provide the reader with additional information about products mentioned in the book. As with any high-technology area, hardware, software, and publications come and go at a fairly rapid rate, but the Resource Section will at least provide a starting point for readers who desire additional information. We have also included a glossary of the basic vocabulary of the field.

In conclusion, the authors wish to stress to the reader that telecommunications is an emerging technology in the sense that the hardware and software are rapidly becoming easier to use and more transparent. We hope you will have patience with telecommunications in its current primitive form. We also hope that this book will convince you of the potential of telecommunications to contribute importantly to the quality of education deliverable to students. We invite you to become part of the growing network of globally linked teachers and students.

1
Telecommunications in Today's World

Tashkent, capital city of the Uzbek Soviet Republic, is located about 250 miles north of Afghanistan. The large auditorium in the Park of Economic Achievement is filled. Among the audience are representatives from the State Committee for Statistics, The Tashkent Building Industry, The Institute of Power Engineering and Automation, The State Committee for Agricultural Production, Aircraft Factory Number 76, The Geological Institute, The Uzbek Ministry of Finance, and The Bureau of Cotton Cleaning. At a long table on stage sits a group of four Americans, representing the communications industry, banks and financial institutions, manufacturing, and education. The Americans are in Tashkent as part of "Information USA," an exhibit sponsored by the United States Information Agency and several U.S. corporations. Designed to introduce American culture and technology to the Soviet people, the exhibit is touring major Soviet cities, remaining in each about six weeks. Tashkent is the third city on the tour. In addition to the display of photocopiers and VCRs, a three-day seminar on some aspect of American industry, science, and technology is held in each city.

It is the last day of the Tashkent seminar. The auditorium is noisy. The Americans are answering questions from the audience. The two Soviet translators struggle to keep up with the pace of the discussion, rapidly changing English into Russian and Russian into English. Several hours go by. Everyone is eagerly talking, laughing, sharing. The topic? Telecommunications.

During the first two days of the Tashkent conference, the Americans addressed the ways that telecommunications is used in the United States to solve communications problems from selling consumer products through telemarketing to helping elementary school students study the problem of acid rain by sharing data through a nationwide telecommunications network. From their questions to the American panelists, it is clear the Soviets recognize that telecommunications is a tool that can help them with their communications needs. Their questions are wide ranging, covering a large number of concerns:

> "How do you get spare parts quickly for tractors on remote farms far away from cities and factories?"
>
> "What are the benefits of optical fibers?"
>
> "Many acres of soil in Uzbekistan are polluted by chemicals from pesticides. We need more information to help us solve this problem. Explain how your scientists are collaborating on this problem."

As the conference in Tashkent indicates, the importance of telecommunications technology is recognized worldwide. It is a tool that can help people find information, communicate with others, and solve mutual problems in all areas including business, industry, government, finance, education, and research.

THE INFORMATION AGE

We are indeed living in "The Information Age." The amount of words, pictures, and numbers we produce each day is staggering. Think of all the information you receive in a single day—from the morning news on the radio as you drive to work, to the advertisements during your favorite television show in the evening. The United States alone has over 9,000 newspapers, 11,000 magazines, 10,000 radio stations, and 1200 commercial and public television stations. Publishers produce over 50,000 book titles annually. The number of computers, all storing information, has skyrocketed from one in 1946 to well over 30 million today. A study in the late 1970s estimated that Americans were exposed to over 8.7 million words a day through newspapers, books, radio, and television. Similarly, the Office of Technology Assessment (OTA) has calculated that American businesses handle over 400 billion documents each year, and that the number is growing by 72 billion documents each year.

Our world contains incredible amounts of information, and the rate at which we produce this information is accelerating. The problem with the staggering amount of information we are creating is the difficulty in gaining access to it all. Take, for example, the field of medicine. Over 3,200 journals are published worldwide each year containing information relevant to health care. How can a doctor, nurse, or other health professional possibly find the information he or she needs on a new drug or surgical procedure? The same problem exists in all fields, from chemistry to law.

We need to be able to use the information we produce. Buried in a mountain of words, pictures, and numbers are answers to questions, help in solving problems, new ideas to explore. We must be able to find what we need quickly and efficiently, and we must bring together information from a variety of sources. For example, a doctor interested in using a new drug therapy must not only be able to locate the latest literature on the treatment, but must also check any possible government regulations restricting the use of the drug. In the past, print technology limited the amount of information we could access and use. Today, even the largest libraries cannot possibly store all the information the world is generating in print form, nor are these libraries necessarily located near the places where people need to use the information. Is there a way out of this dilemma? Telecommunications.

Just as the world is exploding with new information, it is also becoming increasingly complex. Today, answering questions, solving problems, and exploring new ideas requires that people work together. Less and less do individuals work in isolation. Think of the most recent project you undertook. Chances are you worked with other people. A quick survey of several prominent science journals shows that most of the papers they publish are the joint work of several researchers. In business and industry we tend to work in teams, collaborating in the production of an automobile or the design of a new building. This collaboration requires communication, communication with people in the next office, in another city, or around the world. How can we foster communication? Telecommunications.

Increasingly, telecommunications networks are linking the globe, bringing people together from all cultures. As White describes it,

> At a rate and to an extent envisioned only by the most extreme futurists, the world is becoming interconnected by an electronic nervous system over which immense amounts of information flow at nearly the speed of light. (White, 1987, p. 1)

What does the world's growing use and dependency on telecommunications have to do with education? Given that the major goal of education is to prepare our children for their future, then quite a bit. More and more, people are utilizing telecommunications to manage the complexities of the world. We must not only ensure that our children are familiar with this technology, but inspire them to create new and beneficial uses of telecommunications.

THE DEVELOPMENT OF TELECOMMUNICATIONS

The word "telecommunications" simply means communicating across distances. In its broadest sense, this definition includes communicating by any medium including radio, television, telephone, telegraph, and computers. Although other types of electronic telecommunications are extremely important in today's world, this book focuses primarily on computer data communications.

The idea of one computer talking to another arose almost as soon as the first computers were developed in the mid-1940s. By the 1950s, scientists and engineers had created the technology that allowed one computer to communicate with another using telephone lines. However, the actual development of telecommunications networks did not really begin until the following decade.

In 1964, Paul Baran of the Rand Corporation published the first practical design of a telecommunications network, *On Distributed Communications,* as part of a study on communications systems for the Air Force. Around this time, other researchers, including J. C. R. Licklider, Donald Davies, and Lawrence Roberts, were discussing and working on similar ideas. In 1969, the first network, ARPANET (Advanced Research Projects Agency Network), was developed to link four Department of Defense computers.

ARPANET was developed largely as a facility to allow researchers in one part of the country to use computers located in other parts of the country. By using ARPANET and similar telecommunications networks, scientists and engineers could share large and expensive computer systems. For example, an engineer in New Jersey could run programs on a computer located in California. Users of these networks also opened up a new mode of communications, electronic mail—the ability to send messages from one computer to another. Now, not only could scientists in one state run a program on a computer in another state, they also could send messages to their distant colleagues and receive messages back. In fact, it was soon evident that the major use of ARPANET was not to run programs on remote computers, but rather communications via electronic mail among researchers. This ability to communicate easily and quickly with other researchers led to discussions, sharing data, and collaboration on research projects. According to Jennings, *et al.* (1986):

> The major lesson from the ARPANET experience is that information sharing is the key benefit of computer networking. Indeed, it may be argued that many of the advances in computer science and artificial intelligence are the direct result of the enhanced collaboration made possible by ARPANET. (p. 230)

Since the development of ARPANET, the number of commercial and private telecommunications networks has grown rapidly. The first commercial telecommunications networks, TELENET and TYMNET, were available for use by the mid-1970s. The size of the networks has also grown. In 1969 ARPANET linked four computers; in 1971, 23 computers. Today ARPANET connects thousands of computers at universities, industries, and government research agencies throughout the United States. BITNET, formed in 1981, is the world's largest academic network, linking well over 400 colleges and universities throughout the United States with gateways to other academic networks throughout the world.

Today, business and industry are the major users of both private and commercial telecommunications networks. Many major corporations have their own networks. Others use commercial networks such as TELENET and TYMNET to link offices and manufacturing plants scattered across the country or around the world. One large insurance company has a network linking 35,000 terminals to 18 mainframe computers throughout the country. Employees, whether in the same office or in offices hundreds of miles apart, can both use the data stored on these computers and communicate with each other. The network of a major credit card company joins 40,000 workstations, 275,000 point-of-sale terminals, and 17 data centers.

As the number and size of telecommunications networks grows, so do the types of services available on these networks. In a recent talk, Robert Kahn, one of the scientists who created ARPANET, said that creating the first network was relatively simple; the harder part was getting the users to use the network for creative purposes. Early networks let users run a program on a remote computer or use electronic mail. Today telecommunications networks enable people to:

- Send and receive electronic mail from people around the world.
- Access over 3000 databases containing information on everything from the weather to Japanese patents.
- Do library research using bibliographic utilities or online library services such as the Online Computer Library Service (OCLC).
- Run programs on computers, including supercomputers, located thousands of miles away.
- Participate in discussions through bulletin boards and electronic conferences on topics ranging from AIDS to children's television.
- Buy stocks, bonds, and other financial instruments, or purchase consumer goods.

The remainder of this chapter explores examples of how telecommunications is being used in the world today to access information, conduct transactions, and communicate in areas such as finance, marketing, entertainment, and research. The scenarios described here are based on real services and products that are currently available. Although some people may not yet use a computer and modem, it is apparent that telecommunications is already part of daily life and will become increasingly common in our future and our children's future.

BANKING AND OTHER FINANCIAL SERVICES

> *Los Angeles, California.* Three days into your vacation, your cash is getting low. You find a bank with one of the current 45,000 ATMs (Automated Teller Machines), put in your bank card from your local bank in Raleigh, North Carolina, and get an extra $100 from your savings account.
>
> *Rogers, Arkansas.* You are sipping your morning coffee and paying your monthly bills, not using a pen and checkbook, but a computer and modem. Through one of the increasingly available online home banking systems, you quickly make your phone, mortgage, and credit card payments without a trip to the mailbox, and your checkbook is automatically balanced for you.
>
> *Memphis, Tennessee.* Dialing into a brokerage system, you check the latest stock quotes on Wall Street and purchase twenty shares of the hottest stock of the day.
>
> *Salt Lake City, Utah.* Your bank sends an electronic mail message to its corporate accounts each day. The message lets you know the status of your business account and how much money must be deposited that day to cover incoming checks.

Online financial services such as home banking, online brokerage, and automated teller machines are all part of what is called "telefinance." For some years, banks and other financial institutions have been linked by telecommunications networks. Check clearing, credit card approvals, and other financial transactions are all done via networks. For example, each month there are over 50 million credit checks just for VISA card charges alone. All these credit checks are done over telecommunications links among the over 18,000 banks in the VISA system. Similarly, The New York Stock Exchange is supported by a computerized network. Member firms in any part of the country can send orders to buy and sell stock directly to the floor of the exchange.

The use of telecommunications in the world of finance is increasingly important. Recently, Barclays Bank, a major international financial institution, announced plans to spend over $150 million to develop a global telecommunications system that will link bank operations in 2,900 branch bank offices in over 80 countries. As Barclay's director of information services stated, "The bank's strategy is to become one of the top three global financial institutions, and to do that you've got to be networked and you've got to have the whole thing running 'round the clock."

Through online financial services, consumers are able to manage their finances using telecommunications. In addition to letting you buy or sell stocks and other financial instruments online, brokerage services also offer programs that let you track the market, plan your trading strategies, keep records of your transactions, and manage your portfolio. Similarly, home banking services offer a range of financial management programs from balancing your checkbook to computing your net worth.

SHOPPING SERVICES

Wausau, Wisconsin. You need a present for Mother's Day. With your computer and modem, you dial into one of several online shopping services and send Mom a bottle of perfume and a bouquet of roses.

Orono, Maine. You are in the market for a new car. Using an online shopping service, you are able to determine the cars available in your price range, compare the features of different models, arrange a loan, and buy a car. Now, all you have to do is go down to your local car dealer to pick it up.

Shopping online does not completely replace trips to the local mall, but it can be a real convenience for many people. The products available through such services range widely. A recent survey of one such service turned up everything from tulip bulbs to stereo equipment—all "on sale."

APPLICATIONS IN BUSINESS AND INDUSTRY

Herman, Nebraska. A farmer needs a part for his tractor. He telephones his local dealer who finds he doesn't have the part in stock. Using his company's telecommunications system, the dealer checks the inventories of other stores throughout the country, and automatically orders the part from a store in Ames, Iowa. The part is shipped to the farmer that same day.

Boston, Massachusetts. To market a new soft drink, an advertising agency is creating an ad campaign aimed at older adults. Within a few hours of searching several online research databases containing census data and marketing information, the firm has enough data on the buying patterns of people over 30 and competing products to begin designing the campaign.

Prospect, Illinois. Through a network that links the stores of a major drugstore chain throughout most of the United States to a mainframe computer, pharmacists can access a consumer's prescriptions records to find information for tax purposes on how much someone has spent on prescription medicine or allow the consumer to refill a prescription at any store in the chain. With connections to its suppliers, the firm also uses networks to purchase products for its stores.

San Francisco, California. A small software firm needs a brochure to describe its services to potential customers. Using a telecommunications service, it contacts a graphics design firm, works with them to develop a series of "electronic templates," inserts its text into the templates, and quickly produces their brochure.

Hartford, Connecticut. A lawyer handling a bankruptcy case of a wine importer quickly finds information on French bankruptcy statutes and regulations using a large online legal research database.

Houston, Texas. Through telecommunications, an international construction company not only links its offices in several major cities throughout the United States, but employees at these offices can keep in touch with engineers and crews at construction sites in such far-flung locations as South Korea, Egypt, and Great Britain.

Phoenix, Arizona. Every morning the CEO of a fast food chain sits down at his PC and reviews sales in his franchises for the previous day. How many people purchased cheeseburgers in Los Angeles and Pittsburgh? Did more people order French fries with their meal in New York or New Orleans? How is the new salad bar doing in Birmingham? The daily information from point-of-sale terminals at all franchises around the country is transferred, via a telecommunications network, to a central computer every night, so that it is available to top executives the next day.

These scenarios are only a tiny sampling of how telecommunications networks and services are used in business and industry. Accessing information, communications, and handling transactions are all part of the daily use of telecommunications. All areas of business, from research and development to marketing, are taking advantage of the wealth of information available in online databases to increase their competitiveness. Many of these databases are specific to an industry, such as those devoted to information on law, medicine, patents, plastics, paper and printing, chemistry, and automotive engineering, to name just a few.

Telecommunications promotes communications within a company and among corporations. Electronic mail is used regularly by managers and executives to communicate quickly and efficiently, speeding up the decision-making process. Telecommunications also opens up communications between vendors and customers and streamlines ordering, shipping, and handling repairs. Indeed, access to telecommunications is critical to doing business today. According to Bernard Schneider, director of telecommunications for United Stationers, Inc. of Des Plaines, Illinois,

"Maybe five years ago, a network was a strategic advantage. Now if you don't have one, you're not even going to exist in the marketplace."

TRAVEL SERVICES

Jackson Hole, Wyoming. You are planning a trip to Australia. Using one of several online travel planning services, you check flight schedules for over 700 airlines and 3 million flights. After finding the most convenient (and least expensive) flight to Australia, you book your tickets, arrange seating, make a hotel reservation, and reserve a rental car in Sydney. Within a few days, your tickets arrive in the mail.

New York, New York. Employees in a major corporation use an online travel service to book flights and arrange hotel and car reservations. Using the online service lets the company maintain consistent travel policy for employee business trips.

Travel agencies have used networks to make airline reservations for years. Recently, travel services have expanded the types of online information available to both travel agencies and consumers. In addition to making air, hotel, or rental car reservations anywhere in the world, travelers can also access currency exchange information, find out about health requirements, or get the latest travel advisory information from the U.S. State Department.

RESEARCH IN HIGHER EDUCATION

Gettysburg, Pennsylvania. A classics professor finishes the draft of a research paper on Etruscan inscriptions and sends a copy to his co-author in Rome, Italy, using a telecommunications network. After receiving his colleague's comments late that same afternoon, he revises the paper and sends the second draft back to the co-author that evening. Within a few weeks they submit their paper for publication.

Princeton, New Jersey. From her office, a researcher doing work on room temperature superconducting materials accesses a supercomputer located at the Livermore Laboratory in Berkeley, California.

Tulsa, Oklahoma. In fifteen minutes of searching an online database, a professor finds all the citations for the last ten years on the biochemistry of schizophrenia—a search that would take days or weeks using conventional card catalogs and print materials.

Bloomington, Indiana. Every day, college-level classes are broadcast through a statewide fiber optic network to branch campuses and other schools in 17 cities throughout the state. Students no longer have to come to the main campus to take a course for credit. In addition to "attending" lectures, students can also communicate with professors using electronic mail.

Telecommunications is transforming higher education. It is changing how courses and services are delivered to students and also how research is conducted at these institutions. College and university librarians were among the first to see the advantage of networks for sharing bibliographic information. Today, bibliographic utilities such as the Research Libraries Group (RLG), the Research Libraries Information Network (RLIN), the Online Computer Library Center (OCLC), and WLN (Western Library Network) provide students and scholars not only with abstracts and citations, but also with information on which library has a particular book or periodical. A request for an interlibrary loan finishes the search.

Colleges and universities are also using telecommunications to share other resources, including computers and courses. The State University of New York (SUNY) is building EmpireNet, a statewide network that will eventually connect 64 branch campuses and 32 community colleges. Through the network, college-level courses will be broadcast to all locations. According to Charles Blunt, the deputy vice-chancellor of automated information systems at SUNY, "The idea is to overcome geography as a barrier." Students at any of the 96 campuses will have access to all the libraries, computers, and professors of the SUNY system.

Telecommunications is rapidly becoming an integral component of doing research in higher education. Networks such as BITNET, ARPANET, and the new NSFNET link colleges and universities with each other and also with government agencies and other research institutions throughout the United States. These same networks provide gateways to international telecommunications networks and access to researchers throughout the world.

This increased ability to communicate has the potential to change how the work of university researchers is turned into new technologies and products. A recent article in the *New York Times* points out that throughout the twentieth century, the time lag between research and the application of that research is growing shorter. For example, Faraday discovered electromagnetic fields in the late nineteenth century, yet electric generators based on his work were not developed until fifty years later. However, high-temperature superconducting materials were developed only three or four years ago, and already these materials are being used to create new products. Telecommunications has the potential to speed up this process even further by linking researchers in colleges and universities with each other and with their colleagues in government and industry.

PRECOLLEGE EDUCATION

Westwood, Massachusetts. Students talk through an electronic forum with teachers and other adults in the Boston metropolitan area about peer pressure and its relationship to drug and alcohol abuse.

Hilton, New York; Saanich, British Columbia; Cowandilla, South Australia; Elkhart, Indiana; and Queens, New York. Using electronic mail, children from these diverse communities share stories about their towns, favorite books, families, and local customs.

Noatak, Alaska; Maumelle, Arkansas; Holdenfield, Oklahoma; Nashua, New Hampshire; Emo, Ontario; Repulse Bay, Hong Kong; Buenos Aires, Argentina; Haifa, Israel; and 188 other schools throughout the world. Using a telecommunications network, fourth-through sixth-grade students measure the acidity of their rain water and share the data with each other and with scientists studying acid rain.

Oceanside, California, and 26 other schools in the United States, Australia, Germany, France, Canada, and Taiwan. Through an educational bulletin board on a major online database service, 450 students create a database of names and rules of more than 4000 non-commercial games.

Middlebury, Vermont. After a summer workshop on teaching writing at Middlebury College, participating teachers from all over the U.S. continue their discussions using electronic mail.

Telecommunications in precollege education is the subject of the remainder of this book. In subsequent chapters are numerous examples of how telecommunications is being used today to enhance teaching and learning in language arts, science, social studies, geography—in almost all areas of the precollege curriculum and with students at all grade levels, elementary through high school, in both regular and special needs classes. In addition, new ways in which teachers and educators might consider using telecommunications are suggested. Our goal is to stimulate ideas and to help put these ideas into practice in schools and classrooms.

Just as telecommunications is changing how we pay our bills, repair a tractor, or compete in the international marketplace, telecommunications can change education. Telecommunications gives students and teachers access to current, up-to-the minute information, from weather data to the latest discoveries in AIDS research. Telecommunications fosters an interdisciplinary approach to education, connecting science, geography, language arts, social studies, and mathematics in clear and concrete ways. Finally, telecommunications brings students and teachers out of the isolation of their schools and classrooms, promoting collaboration and communication with their colleagues across town, in the next state, and around the world.

REFERENCES

White, M. A. (Ed.). (1987). *What curriculum for the Information Age?* Hillsdale, NJ: Lawrence Erlbaum.

Jennings, D. M., Landwebber, L. H., Fuchs, I. H., Farber, D. J., & Adrion, W. R. (1986). Computer Networking for Scientists. *Science*, Vol. 231, 943–950.

PART I

GETTING ONLINE

2

A Telecommunications Primer

Computer-based telecommunications is the process of connecting two computers together by telephone or cables so that the computers can exchange information in digital form. The ease and efficiency of information exchanged by this method has begun to attract the interest of teachers around the country. Schools and universities everywhere are connecting to public bulletin boards and information utilities to explore the educational potential of teleconferencing, electronic pen pals, and online information retrieval.

Telecommunications requires the use of some special hardware and software. This chapter is designed to get the reader "online" quickly and easily and to explain the more common terms and concepts that are associated with the topic. Chapter 8 gives more advanced technical information. The excitement of this technology comes from using it, and this is the best way to learn. The first part of this chapter describes how to start participating immediately. Given some experience online, information about hardware, software, and host systems capabilities is useful. The second part of this chapter discusses some issues, concepts, and features that are part of the telecommunications world.

GETTING STARTED

In order to use telecommunications, five pieces of equipment are needed:

- a computer
- a disk drive
- a telecommunications software program disk
- a modem
- a telephone line

The computer acts as an information terminal, following the instructions typed at the keyboard and showing the information received over the telephone line on its monitor. The disk drive provides a means of loading the telecommunications software program into the computer and also serves as the location to which received information files are saved and from which prepared files are sent. The telecommunications software program controls the functions of the modem and acts as the interpreter of the commands and information that are sent and received by the computer. The modem acts as the translator for the commands and information going between the computer and the telephone line. Signals coming from the computer terminal to the telephone line are MOdulated into audible sounds, while signals coming from the telephone line to the computer terminal are DEModulated from sounds to computer data bits. The telephone line is the connection to the outside world. Conceivably, a computer can communicate with any other computer in the world, provided a telephone line or other appropriate cable connects them.

In order for the reader to be online as quickly as possible, a typical telecommunications setup is described next. While specific brands of equipment are used here, they are mentioned for illustrative purposes only. Although all telecommunications systems are set up in a similar fashion, users should refer to the manual that comes with the device for its particular installation requirements. When buying these devices, it is wise to have the dealer demonstrate a working system.

Setting Up the Hardware

Among the most popular brands of personal computers found in homes and schools today are the Apple II® (II, II+, IIc, IIe, and IIgs) and the IBM PC® (or IBM compatible). Both families of computers can form the basis of a very sophisticated telecommunications system. These computer systems were also selected as examples because they represent the two main approaches to setting up telecommunications: manufacturer based, as we will illustrate with an Apple computer system; and product based, as illustrated with an IBM system.

Manufacturer-Based Telecommunications

Apple Computer Inc. supplies computers, disk drives, Apple® Personal Modems, and Apple Access II® telecommunications software that are specifically made to work together. Please note that Apple® computers are compatible with a wide range of disk drives, modems, and communications programs (including the Hayes Smartmodem used in the IBM PC system we describe further on in this chapter), even though the example system used here is composed entirely of Apple manufactured equipment and software. Such a system represents a manufacturer-based approach to setting up a telecommunications system. The advantage of this approach is the virtual certainty that the components will work together compatibly and the convenience of dealing with a single service and technical department in the event of problems. The drawback is the limited choice of components and features.

For an Apple telecommunications system, the basic computer equipment needed consists of the Apple IIe® with 128k of memory, an Apple II® monochrome monitor, and a $5\frac{1}{4}''$ disk drive. A Super Serial Interface Card with cable is necessary to make the connection between the computer and the Apple Personal Modem. Following the manufacturer's instructions, and with *the power turned off,* install the Super Serial Card in expansion

FIGURE 2.1. Super Serial Card

slot #2. The expansion slots are found along the inside back of the computer and are accessed by lifting the top of the computer upwards from the back. Be sure that the "jumper block" and the two sets of switches on the Super Serial Card have been set to the correct positions for modem operation as shown in the Super Serial Card's user manual.

The Apple Personal Modem is designed with a self-contained power supply that plugs into any three-prong grounded electrical outlet. An audible tone sounds for several seconds when the modem is plugged in to indicate that it is functioning properly. Along its bottom edge are three jacks and a dial. Two of the jacks are standard modular telephone connectors. To install the modem, begin by disconnecting your telephone's modular telephone cord from the modular wall jack. If the wall jack is not modular, see the section in this chapter, "Telephone Hookup" or consult the modem manual for instructions on how to modify the jack for modem use. Connect the modem to the telephone line with the modular telephone cable that is provided with the modem by plugging one end into the modular wall jack and the other into one of the modular jacks on the modem. In order to be able to use the telephone in the usual way when not telecommunicating, plug its modular cord into the other modular jack on the modem. Finally, use the 8-pin cable purchased along with the Super

FIGURE 2.2. Apple Telecommunications System

Serial Card to connect the modem to the interface connection just installed in the back of the computer. The computer hardware is now ready for telecommunications.

Product-Based Telecommunications

A product-based telecommunications system is comprised of a set of components produced by two or more different manufacturers. Each of the manufacturers designs components that perform one or more of the telecommunication system's functions in conjunction with one or more products made by other manufacturers. It is the responsibility of the buyer to determine whether any particular component is supposed to work with the others in the system. The advantage of this approach is the wide choice of components that permits the buyer to put together a customized telecommunications system providing the features the buyer most needs or desires. The drawback is the uncertainty that the components will work together compatibly and the need to deal with multiple service and technical departments in the event of problems.

A telecommunications system comprised of an IBM PC with 512k memory and $5\frac{1}{4}''$ disk drive, IBM compatible Serial Communications Card,

FIGURE 2.3. Serial Communications Card

Hayes Smartmodem 1200, and Procom Plus telecommunications software is an example of the product-based approach. Following the Serial Communications Card manufacturer's instructions, and with *the power turned off*, install the card in an expansion slot in the IBM PC®. The expansion slots are found along the inside back of the computer and are accessed by sliding the computer out of its case. Be sure that the switches on the Serial Communications Card have been set to the correct positions for modem operation as shown in the card's user manual.

The Hayes Smartmodem 1200 is designed with a self-contained power supply that plugs into any three-prong grounded electrical outlet. A red LED (Light Emitting Diode) lights up on the front panel to indicate that the modem is functioning properly when it is plugged in and turned on. Along the back edge are two modular telephone jacks. To install the modem, begin by disconnecting the telephone's modular telephone cord from the modular wall jack. If it is not modular, see the section in this chapter, "Telephone Hookup" or consult the modem manual for instructions on how to modify the wall jack for modem use. Connect the modem with the modular telephone cable that is provided by plugging one end into the modular wall jack and the other into one of the jacks on the modem. In order to use the telephone in the usual way when not telecom-

FIGURE 2.4. IBM PC Telecommunications System

municating, plug its telephone cord into the other modular jack on the modem. Finally, use the cable purchased along with the Serial Communications Card to connect the modem to the interface connection just installed in the back of the computer. The computer hardware is now ready for telecommunications.

Telephone Company Notification

When the assembly of the telecommunications system is completed, notify the local telephone company business office that a modem is being connected to its lines. The phone company might request the following information, generally found in the modem's user manual:

- FCC registration number
- Ringer equivalence number
- Modem manufacturer's name
- Modem model number

Setting Up the Software

A microcomputer like the Apple IIe or IBM PC can do many things. What it does at any given time depends on the program that has been loaded into the memory. To use the computer for telecommunications, place the $5\frac{1}{4}''$ floppy disk containing the program (Apple Access II for the Apple IIe® system; Procom Plus for the IBM PC system) into the computer's disk drive and turn on the computer. The computer loads the program into its memory and runs it. This process is known as "booting," and is the same for all microcomputers. (*Note:* Apple and IBM PC computers cannot read or use one another's software.)

Configuring the Apple IIe Telecommunications Software

The computer that is being called, known as a host computer, has certain configuration requirements for establishing communication. Both computers' telecommunications programs must be set compatibly so that the two systems can "talk" to each other. Apple Access II provides this configuration capability from the main menu that appears on the monitor when the telecommunications program has been loaded.

- Choose 3, Set Up Communications.

From the Set Up Communications menu, set the following configuration:

- Choose 1, Terminal Characteristics

Using the number keys to select categories and the tab key to set conditions, specify the following:

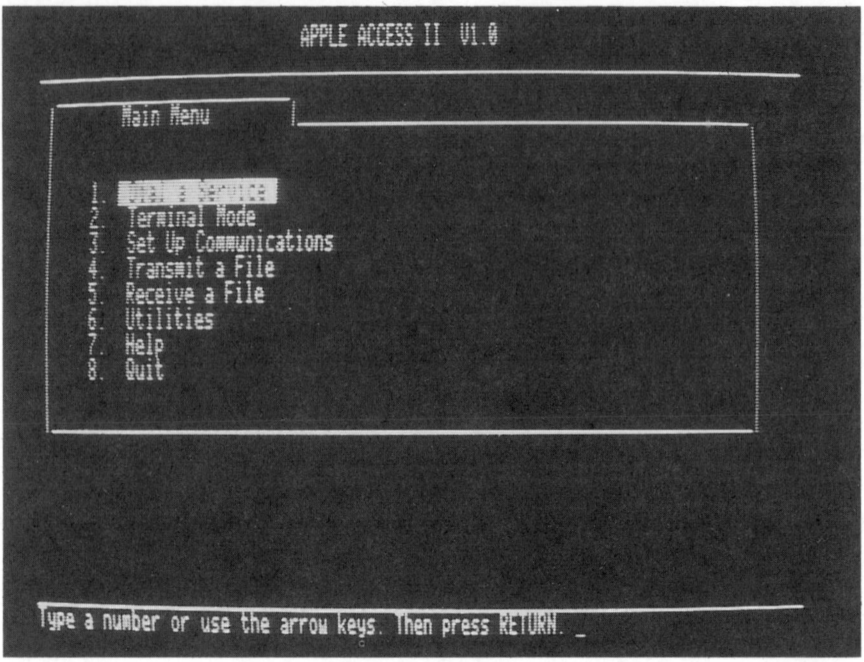

APPLE ACCESS II V1.0

Main Menu

1. Dial a Service
2. Terminal Mode
3. Set Up Communications
4. Transmit a File
5. Receive a File
6. Utilities
7. Help
8. Quit

Type a number or use the arrow keys. Then press RETURN. _

FIGURE 2.5. Main Menu

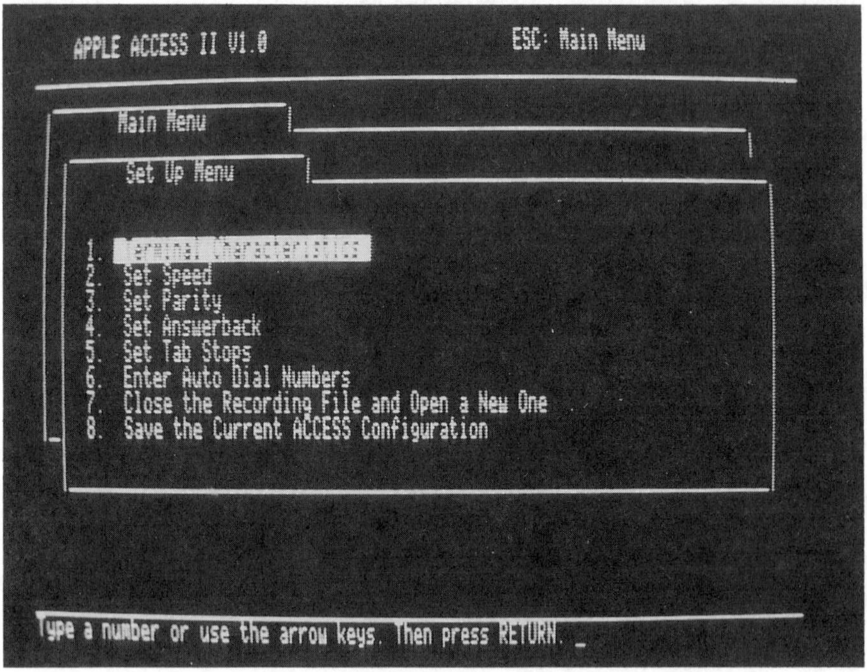

APPLE ACCESS II V1.0 ESC: Main Menu

Main Menu

Set Up Menu

1. Terminal Characteristics
2. Set Speed
3. Set Parity
4. Set Answerback
5. Set Tab Stops
6. Enter Auto Dial Numbers
7. Close the Recording File and Open a New One
8. Save the Current ACCESS Configuration

Type a number or use the arrow keys. Then press RETURN. _

FIGURE 2.6. Set-Up Menu

1. TTY
2. Do NOT send LF after CR
3. 8 bits per Character
4. Enable XON/XOFF
5. Normal Video
6. Full Duplex
7. Wraparound

Press the ESC key to return to Set Up menu.

- Choose 2, Set Speed.
  Press the number 2 key to select 300.
  Press the ESC key to accept and return to Set Up menu.
- Choose 3, Parity.
  Press the number 1 key to select NONE.
  Press the ESC key to accept and return to Set Up menu.
- Choose 8, Save the Current ACCESS Configuration.
  Press the RETURN key to save the setup and return to the Main Menu.

Configuring the IBM PC Telecommunications Software

When the IBM PC begins running Procom Plus, the program goes directly into terminal mode with a status bar appearing across the bottom line of the computer monitor.

Just as is the case with the Apple II system, the computer being called, known as the host computer, has certain configuration requirements for establishing communication. The IBM PC telecommunications software must be set compatibly if the two systems are to "talk." The status bar indicates the software configuration settings and is set up as follows:

Press the Alt and Z keys simultaneously for the Main Menu.
Press the Alt and S keys simultaneously for the Set Up Menu.
  Select Terminal Options by pressing the down arrow key.
  Select General Options by pressing the down arrow key.
    Press the A key until Terminal Emulation reads TTY.
    Press the B key until Duplex reads Full.
    Press the C key until XON/XOFF reads On.
    Press the E key until Line Wrap reads On.
    Press the F key until Screen Scroll reads On.
    Press the ESC key to return to General Options.
  Select Setting Options by pressing the down arrow key.

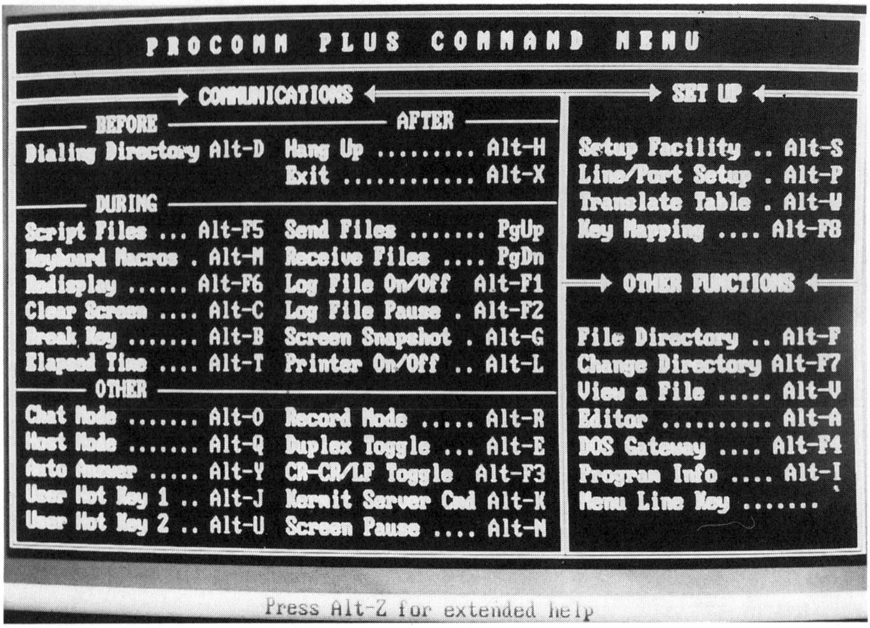

FIGURE 2.7. Status Bar

Press the appropriate keys to set:

Baud to 300

Data Bits to 8

Parity to None (N)

Stop Bits to 1

Press ESC to return to the Setting Options.

Press ESC to return to Main Menu.

Press ESC to return to Terminal Mode.

Testing the Modem

The computer telecommunications systems are now configured to act as though they are teletype machines with an "8N1" setting at 300 baud. This is one of the most common telecommunications settings, the one that is generally used unless the host computer is known to require something different (1200 baud is becoming more and more common). The meaning of these configuration settings is discussed in Chapter 8.

Three simple tests determine whether the computer and modem have been set up properly and are ready for telecommunications. If these tests indicate that the system is not functioning, and the suggestions for fixing the problem do not work, refer to the "Troubleshooting" section at the end of this chapter or consult the equipment dealer.

First, the Apple Personal Modem "beeps" when it is plugged into an active electrical outlet, while the Hayes Smartmodem 1200 has a power lamp that lights up when electricity is applied. If the modem does not give the proper indication, check to see that the power cord and plugs are firmly seated in their sockets. If the modem still does not work, check to see that the outlet is active by plugging in something else, such as a reading lamp. If the outlet is not working, it might be controlled by an on/off switch located on the wall or on a power strip. Finally, check the fuses or circuit breakers.

Second, lift the receiver of the telephone that is plugged into the modem. A dial tone indicates that the phone line is working. If there is no dial tone, check all the modular jacks again and reseat any loose connections. If there is still no dial tone, plug the telephone back into the wall jack to verify a dial tone. If there is none, the problem is in the phone line. If there is a dial tone, test the telephone cord that came with the modem by plugging it in between the telephone and wall jack. Replace the telephone cord if it appears defective.

Finally, enter Terminal Mode (Apple Access II Main Menu choice 2; boot-up condition of Procom Plus) and type the keys A, T, RETURN. This is the Hayes Smartmodem command set "AT" or "attention" command, which is recognized by both the Smartmodem 1200 and the Apple Personal Modem (and most other "Hayes compatible" modems on the market today). If your modem is not "Hayes compatible," check the owner's manual that came with it for the equivalent attention command. If everything is connected properly, the "OK" reply will be displayed on the monitor screen. If not, check to see that baud rate is set for 300 or 1200 bits per second.

If none of this solves the problem, do not be discouraged. Just go back over the installation process and check that all circuit cards, cables, and connecting jacks are correctly located and securely seated, and that the communications software is configured correctly for the equipment. Most problems turn out to be caused by missing or loose cables or switches that are off. When the system has been properly set up, the "AT" command returns the "OK" message on the screen. This indicates the readiness of the system to enter the exciting world of telecommunications.

A Telecommunications Tutorial

To enter the world of telecommunications, you must "log on" to a host computer, a process that is quite standard and easy to do. All microcomputer telecommunications systems function in the same manner once they have been set up and the software configured. From this point on, there is no difference between the operation of the Apple IIe and IBM PC systems. The main thing to remember is that the computers operate the telephone system for telecommunications purposes in the same way that people do when they make telephone calls. Nothing is different about the way calls are made. As a result, keep in mind that telephone usage charges are the same for both local and long-distance calls whether a person or computer is calling.

As a demonstration of a typical telecommunications session, what follows is a listing of a "guest" visit to LES-COM-net, a bulletin board supported by Lesley College of Cambridge, Massachusetts, and frequently used by the authors. Bulletin boards are sometimes referred to as a BBS, for Bulletin Board System. This BBS is an "open board," meaning that membership is not required before access is permitted. Try calling this bulletin board by following the procedure below. Note that this is a long-distance call from outside the 303 area code (northern Colorado). As an alternative, find the number of a local computer bulletin board in your area. These phone numbers can generally be obtained from a computer store. Often stores have notice boards where people post want ads and BBS numbers. Most BBS systems use a similar log-on procedure. The following is a description of the general process.

Making the Telephone Call

To make the telephone call, enter the Terminal Mode (Apple Access II main menu choice 2; boot condition for Procom Plus on the IBM PC). The computer clears the screen and is ready to instruct the modem to make the telephone call. Typing a series of keys (called a command string) gives instructions to the modem.

The Hayes dialing command strings, used by the Hayes Smartmodem 1200 and Apple Personal Modem, begin with the letters AT D meaning "attention and dial."

With touchtone telephone service, type: AT DT 1 303 526-2046
With rotary dial service, type: AT DP 1 303 526-2046

Listen as the telephone call is being made. Both modems have a built-in speaker that lets the user listen to the call. When connection with the host computer is established, the following information is displayed on the computer screen.

```
AT DT 1 303 526-2046
CONNECT
WELCOME to the Lesley College Communications Network
    (303) 526-2046
  300/1200/2400 baud-24 hrs/day
    Welcome EXfer Sysops!
```

```
You are connected to phone line #1!!!!!
New users type "NEW"
```

Online Activities

Congratulations! You have just made your first telecommunications contact! Let's explore the bulletin board. Since this is your first call to the bulletin board, you are considered a "new" user, so respond to the Account Number prompt by typing the word NEW, followed by a carriage return.

```
Account Number? (User Number)
-->NEW

Enter your real full name. [20 chars max]
:YOUR NAME

City? [16 chars max]
:YOUR CITY

State? [Form: XX]
:YOUR STATE

Phone Number? [Form: ###-###-####]
:YOUR PHONE NUMBER

We have the following.

YOUR NAME
YOUR CITY, STATE
YOUR PHONE NUMBER

Is this correct? (Y/[N]):Y
```

At this point there are two options: request a password and account number to become a member of the board, or choose to be a guest. For now, choose to be a guest. This allows you to look around and sample what the board has to offer. To become a regular user after looking around the board, a main menu choice exists that requests that a password and account number be established. If you decide to join the board after ending this session, you will need to repeat the above procedure. For now, type G for guest access.

```
-->G

Good morning YOUR NAME,
It's 6:11 AM on 07/12/89
Connected at [300] baud!
You were last on UNKNOWN
Time left today: 19 minutes.

Ctrl-S Stop/Ctrl-Q Start Spacebar to Exit

Welcome to ->LES-COM-net
```

Sysop ------>George Willett

Sponsor ----> Lesley College

[18][Main Level] Option (?=Help):?

Let's ask the bulletin board for "help" by typing a ? at the prompt. Help is a feature found on many systems. It provides more information about the choices or how to respond to prompts.

<LES-COM-net>
<Main Command Menu>

D -> Define system display params
S -> Examine your Status
P -> Receive a password to log-on
W-> A-WALK around LES-COM-net
J -> How to JOIN LES-COM-net

C -> Chat with the sysop
F -> Feedback to Sysop
E -> Read your E-Mail

H -> Get detailed help
I -> System information
LN-> LES-COM-net news

T -> Terminate session

Let's take the guided tour of the bulletin board. By choosing main menu choice W, the system explains its main features.

[17][Main Level] Option (?=Help):W

Ctrl-S Stop/Ctrl-Q Start Spacebar to Exit

****

WELCOME TO LES-COM-net

This little tour is screen formatted for 80 columns. If you can't display 80 columns, this program will be a little confusing. Sorry.

LES-COM-net has many menus and boards from which to choose. The first menu is for guests or members who have not yet received a validated password. The section marked H will give you all the commands you need to learn to communicate effectively. Some of the commands in the detailed help section may not be available to you if you are a limited access user, but most will be available when you receive a validated password.

When you become a validated user, you will be given partial access to the system and the menu you will see when you log on will look like the following:

<LES-COM-net>

<Main Command Menu>

X -> EXfer Sysop Area ! H -> Get detailed help
B -> Bulletin Board Menu ! I -> System information

B1 -> Bulletin Board #1
LN-> LES-COM-net news
LT-> Lesley Telcom class
$ -> Paid Members Menu
D -> Define system display params
S -> Examine your Status
P -> Change your password
W -> A-WALK around LES-COM-net
J -> to JOIN LES-COM-net
V -> Vote on the current topic

! O -> Other BBS numbers
! 01-> More BBS numbers
! 02-> Local BBS numbers
! U -> Get a user listing

! C -> Chat with the sysop
! F -> Send feedback to sysop

! E -> Send/Read your E-Mail

T -> Terminate session

There is no charge for this limited system and as long as you call at least once a month you will keep your validated password. At this level you can use the General Bulletin Board and send and receive mail.

EXfer sysops have access to the EXfer sysop area from this menu. As a member you are encouraged to use our mail and bulletin board service for your communication convenience.

You may make requests, leave information for people, and even run advertisements.

When you press B from the main menu you will get the following menu. Remember you will have access to only B1 and B2 as a non-paid member.

LES-COM-net Bulletin Boards

All LES-COM-net Members
B1 -> General Bulletin Board
B2 -> Odyssey of the Mind
Paid LES-COM-net Members
B3 -> Basic Programming Board
B4 -> Pascal Programming Board
B5 -> Logo Programming Board
Special Boards
B6 -> Western Status Comp. Coor.
B7 -> Lesley Faculty Board

A bulletin on B1 goes something like this:

FROM ->name of posting user (user #)
DATE ->mm/dd/yy

Text of message.
This could be a general announcement, a
question that you would like answered,
a response to another bulletin, or your
thoughts on one of the topics of discussion
on the board, to name a few of the possibilities.

Whatever is posted will be seen and read by all
users with access to this board.

The EXfer transfer section (EX from the main menu), for PAID members only, has fourteen libraries and, as you can see from the menu below, contains many useful programs.

[001]: General-Applesoft BASIC 3.3
[002]: General-Applesoft BASIC Pro
[003]: LESLEY PROJECTS DOS 3.3
[004]: Lesley Projects Pro DOS
[005]: Apple][gs Pro-DOS
[006]: General-Terrapin LOGO
[007]: Lesley Projects-Terrapin Logo
[008]: Pro-DOS Utilities
[009]: Lesley Projects-Instant Pascal
[010]: Macintosh Programs
[011]: Upload to SYSOP!
[012]: Private Section (EXsym Programmers)
[013]: IBM Public Domain
[014]: Computer TXT Pictures

LES-COM-net will continue to add quality downloads as we strive to become the finest public domain library in the United States.

The General files section contains more useful information. It is not unlike the download section, but most of the files are shorter and can be useful by just viewing them.

You can also copy these files into your modem buffer if you like.

We hope you enjoyed your tour and that it gave you a little insight into the information available to you on LES-COM-net. You are welcome to leave FEEDBACK TO SYSOP if you have any questions or comments. Remember, to receive FULL validation you must send $10.00 along with the information requested in the introduction to:

LES-COM-net
2250 Juniper Court
Golden, CO 80401

LES-COM-net is independently operated in cooperation with:

Lesley College
29 Everett St.
Cambridge, MA 02238
(617) 868-9600
1-800-999-1959

AND

Professional Outreach Associates
710 11th Ave. Suite L90
Greeley, CO 80631
(303) 353-6502
(303) 654-1037 (Denver Phone Number)
****

This BBS is a menu-driven system. Options are displayed following the completion of each routine used. Now explore some of the other menu choices on the "guest" board. Simply type the letter corresponding to your choice at the [Main Level] Option: prompt. When you have finished, respond with the T option to terminate, or end, the online session. When you do this, the log-off process appears as follows:

Logging off the BBS

```
[07][Main Level] Option (?=Help):T
Terminate Connection! . . . Are you sure? ([Y]/N):Y
Goodbye YOUR NAME, you were caller #21447
Thank you for calling LES-COM-net.
Connected nn mins, nn secs
GBBS 'Pro' Copyright 1986 by L&L Productions V : 1.4b
NO CARRIER
```

In this example online session, communications are concluded in the normal log-off fashion. There are circumstances, such as when the host system inexplicably ceases communications, when it is also useful to know an alternative way to end a telecommunications connection. The following procedure commands the modem to hang up, or break the telephone connection. When this happens, the host computer (located in Colorado in this example) detects that the telephone connection has ended and automatically logs off. The main concern in such a situation is to be sure that hanging up the telephone has been successful so that the connection between the modems is broken. Simply leaving terminal mode or turning off the computer does not automatically end the telephone connection. Once two modems have established contact with each other, only the loss of a carrier signal causes them to hang up the telephone. This occurs only when the host computer logs off the user, the call is interrupted somewhere between the two modems, or a "hang-up" command is issued to the modem. With Hayes compatible modems, such as the Hayes Smartmodem 1200 or Apple Personal Modem, this so called "panic button" process is as follows:

1. Press +++ (the "plus" key three times in rapid succession). The modem prints OK on the screen to indicate that it is in the command mode.
2. Now type AT H to command the modem to hang up. The modem prints OK on the screen to indicate that it has hung up the telephone (pressing Alt and H keys simultaneously with Procom Plus accomplishes the same thing).
3. Verify that the computer is indeed offline either by listening for the dial tone if there is a telephone connected to the modem, or by typing AT O to reenter the modem's data mode. When attempting to reenter the data mode after an online session has been ended, the modem will print the "No Carrier" message on the screen if the telephone line has been successfully disconnected.

Returning to the Main Menu

At this point, the computer is offline (no telephone connection exists), but is still in terminal mode. To make another telephone call, just use the AT D command as done earlier. If the telecommunications session is over, return to the Main Menu of the Apple Access communications program by pressing the open-apple and ESC keys at the same time, or press the Alt and Z keys in Procom Plus, and select QUIT. In fact, it is quite possible to return to the Main Menus this way at any time when in terminal mode. Be careful in doing so while still online, because this does not end the telephone connection, so long distance and access charges might still be applied.

INTERMEDIATE TELECOMMUNICATIONS SKILLS

Telephone Hookup

All direct-connect or in-line modems on the market today use modular connectors that are designed to attach to the telephone line in place of the telephone set. This means that the telephone cord fits into a "jack" with a small plastic connector that has four wire contacts exposed on one side. This design facilitates the easy movement and exchange of one telephone with another.

Older telephone installations may use plugs or be "hardwired." The plug design is a forerunner of today's modular connector and can be distinguished by its one-inch-square size and four round legs that fit into a similarly designed receptacle. Hardwired means that the telephone is connected directly to the outside line and was never intended to be disconnected or moved. Using a modem with these types of telephone installations requires that they be converted to a modular form. This can be done either by calling the local telephone company to do the job, or by doing it with parts obtained from a telephone supplier.

FIGURE 2.8. Plug Conversion

To convert a four-prong telephone jack for modem use, obtain a "four-prong to modular adapter" for the wall jack and a "modular to four-prong adapter" for the telephone. To convert a hardwired telephone jack for modem use, obtain a "modular jack converter" for the wall connection.

FIGURE 2.9. Hardwire Conversion

Telephone Service

The use of a personal computer and modem for telecommunications generally assumes that the system is connected to a single-user telephone line. This is sometimes called a "direct" line and it means there is a unique telephone number for calling the line to which the telephone connects. Simple, personal telecommunications is difficult, if not impossible, with other types of telephone line arrangements. In such cases, consider the installation of another telephone line for use by the telecommunications system. The phone company may well suggest a "private data line." However, an ordinary voice-quality line does just fine without a substantial extra expense.

Extension phones are a problem because if someone attempts to use one while a telecommunications session is in progress, he or she interferes with and garbles the signal, at best, or interrupts the carrier signal and causes the modem to hang up, at worst. When a telephone line is being used for telecommunications, it cannot be used for voice calls at the same time.

Party-line telephones have the same problems that extension phones do, with the added danger that during telecommunications an interrupting party cannot indicate when they have an emergency situation requiring that the telephone line be surrendered. Party lines also are hard to use for autoanswer telecommunications, such as setting up a BBS, because the other party might answer the telephone before the modem does. Besides, the call might be for the other party.

Multiple-line phones, found frequently in small offices and schools, cannot be used "as is" for telecommunications because they do not have the standard four-wire modular connectors. Their jacks contain extra wiring to control which extension phone is active and are not compatible with standard modular modem connectors. This problem can be overcome by asking the local telephone company to install a "line splitter" modular jack for one of the lines to provide access for telecommunications.

Switchboards may or may not pose problems, depending on their design. Usually, outgoing telecommunications calls can be made through a switchboard. Manually operated switchboards require speaking with an operator to get a dial tone before proceeding with the telecommunications call. This limits telecommunications use to those hours when an operator is on duty. As with extension phones, there is the potential for interference from the operator that would garble or break the modem connection. Automatic switchboards that access outside telephone lines by dialing a special code can generally be used for telecommunications since most communications software can make extended code calls, including the necessary pauses that allow the switching systems time to work. The primary limitation on telecommunications in these cases is whether long-distance calls are needed and permitted. Many automatic switchboards only permit local calls without operator assistance. Also, only the most sophisticated of automatic switchboards permit an outside caller to contact a specific extension without operator assistance. Calling in through a switchboard is generally limited to making connections only when the other party is present and both have communications software that can be used in "chat" mode (discussed in Chapter 8).

Even when using a direct line for telecommunications, it is possible to experience problems. The most common of these is disconnection caused by "call waiting." With the new computer-controlled telephone switching offices, more and better telephone services are being created all the time. Any service that interrupts a call in progress causes problems for telecommunications. At the present time, call waiting is the most common user service that interrupts a call in progress. To avoid having a telecommunications session cut short, this service should be disabled. This may be as easy as dialing a special code to temporarily stop call waiting, or the service may have to be discontinued entirely.

Using More Advanced Features

Familiarity with the set-up and log-on process of telecommunications usually leads to a desire to do more than just read and post messages. In fact, with a 1200 baud modem, the information is displayed on the screen more quickly than can be read. Data that scrolls off the top of the screen might be lost forever. Saving incoming messages for later use is therefore quite important. The ability to save information even after it scrolls off the screen depends on the communications software. Fortunately, all but the simplest of terminal programs provide several means of saving data. The two most common methods are the online printer option and the "capture" or save buffer.

Online Printing

Online printing means that everything that appears on the screen is sent to a printer at the same time. This provides a hard copy or printed record of the online session, provided the printer is turned on. With Apple Access communications software, experiment with this feature by pressing the open-apple and P keys at the same time; with Procom Plus, press the Alt and P keys at the same time. This toggles (switches) the printer to the online mode. Pressing both the open-apple and P or Alt and P keys again stops the data from going to the printer. For other communications software programs, look in the owner's manual index for the section on printing to find out how to activate the online printer mode.

Capture (Save) Buffer

An even more versatile method of saving information is the capture or save buffer (called the *Record File* in Apple Access II, and the *Log* in Procom Plus). This feature uses any extra memory not needed by the communications program itself to store what appears on the screen. Generally, the contents of the buffer can be printed offline at a later time. The advantage of this is that some printers cannot run at 1200 or 2400 baud and hence slow down the telecommunications process when used for online printing at those speeds.

The option of saving the buffer as a disk file also is usually available. This feature makes use of the power of the computer to collect information in digital form for processing and analysis with other application programs. Keep in mind that a capture buffer only saves ASCII character (sometimes called text or TXT) data. It is therefore very well suited for all forms of messages, electronic mail, bulletins, and library type information. It can be used for any other type of data that has been converted to ASCII form. A capture buffer does not employ any form of error checking during the transfer process. Inspect the received message to make sure that no errors occurred.

One of the most common uses of this feature is to save a message or bulletin for use in a word processor. This allows the text to be cleaned up, by removing unnecessary bulletin board command messages and correcting any errors caused by noise on the telephone lines during the transmission process, before using the material. With the Apple Access communications program, the capture buffer is toggled on by pressing open-apple and R keys at the same time; with Procom Plus, the Alt and L keys are used. To indicate that the capture buffer is active, the cursor begins to blink on and off in Apple Access II, while the status bar indicates that the Log is on in Procom Plus. Now proceed with the online session just as before. The only thing that is different is that everything appearing on the screen is not being lost when it scrolls off the screen. It is being saved in the capture buffer and is available to read again later. When everything is saved, or the online session is completed, pressing open-apple and R or the Alt and L keys again turns off the capture buffer.

While using the capture buffer with Apple Access II, the disk drive may be active on one or more occasions, indicating that data is being saved to disk. If you have not created, or been prompted for a file name for the recorded file, it is on the Apple Access disk as the filename TERMREC (Terminal Record). If a different name is more appropriate for the file, use the *Set Up Communications* option from the Main Menu (remember: press open-apple and ESC keys together to exit from the terminal mode back to the Main Menu) and select number 7—*Close Recording File and Open a New One*. A prompt appears for the filename. If more disk space is required for the file, use the *Utilities* option from the Main Menu and select number 2—*Format a Disk* to create a special save disk. To save the capture buffer in Procom Plus, press the Alt and Z keys to get the Main Menu and choose the file management menu options to perform the same disk management tasks.

Using Files with Word Processors

ASCII files created by saving the capture buffer to disk are compatible with most word processing software programs. This compatibility arises from the common use of ASCII text (TXT) type of files by both types of programs. Most word processors, therefore, have an option of loading an ASCII text (TXT) file from disk. In actual practice, the ability to do this assumes that both the word processing program and the communications software use the same disk file management system. Determine this by finding out what "DOS" (disk operating system) is used by the word processor and checking to see whether the communications software is compatible with it. Unless the telecommunications file has been saved with the same DOS used by the word processor, it may not be readable by the word processor's file loading command even though it is in the ASCII text (TXT) format. Apple Access II is a PRODOS based system; Procom Plus is MS-DOS® based.

Appleworks is an example of a word processing program that is compatible with Apple Access ASCII text files. To move a saved capture buffer file into Appleworks as a word processor file, start up the Appleworks program, select the *Add Files to the Desktop* Main Menu choice, the *New File for the Word Processor* option, and the *Make a New File from a Text (ASCII) File* as the source. Put the Apple Access disk back in the disk drive and the word processor loads the ASCII source file. It can then be edited, changed, and saved as a word processor document. A similar procedure permits an MS-DOS word processing program—WordStar, for example—to retrieve Procom Plus capture buffer ASCII text files.

A communications program or word processor that allows saving the capture buffer in either line or paragraph modes makes this task easier. The telecommunications process adds a carriage return character to the end of every line it sends and if the software cannot filter these out, only line-oriented text appears in the word processor. Also consider communications software that offers built-in word processing as one of its features.

Sending ASCII Files

The ASCII compatibility of word processor and telecommunications programs also makes it easy to send text (TXT) files created prior to an online session. Most communications software sends ASCII text (TXT) files on command. With Apple Access, the command is issued by pressing open-apple and C keys at the same time; the Alt and Z keys bring up the Procom Plus main menu from which this option can be selected. In both cases, the software then prompts for the name of the file to be sent.

The two main advantages of this ability are offline editing and speed of transmission. Offline editing means composing and editing messages without tying up the telephone lines. The work can be done at any time and the text sent when it is convenient. While online, the computer can "type" the messages much faster than you can. At a speed of 300 baud, the computer is "typing" at a rate of 30 characters per second or about 360 words per minute. This cuts down on the online time, that both conserves the resource for others and saves money in long-distance or connect-time charges.

Uploading and Downloading Files

For error-free ASCII file transfer or exchanging program listings, data, or code that is in binary format, some form of file transfer protocol is needed. The process of using a transfer protocol is known as "uploading" when sending the file and "downloading" when receiving it. Although ASCII file transfer is considered to be a transfer protocol in the strict sense of the word, the term usually refers to Xmodem, Ymodem, or Kermit, explained in more detail in Chapter 8. Specific instructions on the use of these protocols is dependent on the communications software and the implementation of the protocol on the host system. Consult the communications software manual and host system operator for exact instructions for error-free file transfer.

Troubleshooting

Fortunately, most telecommunications systems work as expected as soon as they are set up. If difficulties arise with the telecommunications system, however, try to identify the exact nature of the problem. If it turns out that technical assistance is needed to correct the problem, the information provided from your investigations may well save both time and money on the repair. It is always a good idea to check the simple and most obvious things first. Repeat both the hardware and software installation process. Unplug and reconnect all cables and interface cards, tightening any screws to ensure good connection. This procedure can cure many problems. Computer connections tend to develop a layer of oxidation that can eventually break the electrical path between two devices. Reseating the plugs clears the oxidation away and reestablishes the electrical pathway.

If this does not cure the problem, the following approach might help locate the source of difficulty:

1. *Identify the problem.* Determine the exact symptoms and define the nature of the problem.
2. *Isolate the problem.* Determine the source of difficulty. Identify the last correct action before the problem occurs. Experiment with possible variables—baud rate, parity, number dialed, cable used—one at a time until the system acts differently.
3. *Confirm the problem.* Test the hypothesis to see if the problem can be made to occur regularly and reliably.
4. *Correct the problem.* Fix or replace the necessary components.

Some common sources of problems and their cures follow.

Garbled information displayed on the screen.

If the garbled information consists of one or a small number of characters, and they occur in sporadic, random fashion, it is a common occurrence and nothing to worry about. This is generally caused by noise and static on the telephone line. If the problem is severe, hang up and place another call to get a quieter connection. If the entire message is unreadable, check the communications software configuration to make sure that it is compatible with the host system's. Check especially for baud rate, parity, and duplex settings. If the configuration is correct, some other possible sources of problems are:

- loose data cables: Check all connections.
- telephone handset off hook: Noise entering line.
- extension phone: Someone listening in or trying to use the line.

No information on the screen.

Incorrect baud rate, duplex, or keyboard echo setting results in no display of either one or both of the computer's information exchanges. Check for compatibility with the host system and change the configuration settings if necessary.

Regularly missing portions of the information exchange.

Check to see if the communications software supports the use of Xon/Xoff information flow control and activate it if it does. If the missing information always occurs at the beginning of each new line, set transmission nulls in the output protocol of the host system (see Chapter 8 for information on nulls).

Unable to establish communications.

Wrong phone number for host system. Place a manual call and listen for a high-pitched carrier tone. If a person answers, ask if this is the correct number for the host system, as some only operate after hours or on weekends.

Incorrect baud rate or parity settings. Experiment with these to find the right combination, or call the host system operator or consult its operating manual for the correct settings.

Frequent interruption of communications.

The call waiting service may be interrupting the modem connection. Turn off or cancel call waiting, or add call forwarding and route calls to another phone.

The host system may be having problems. Call the system operator using voice to inquire.

The telephone line may be causing problems. If the same problem occurs with voice calls, the telephone line is the problem. If not, the modem may need service.

As mentioned before, most problems turn out to be simple and easily solved if diagnosis is approached in a careful, organized manner. Another effective approach is to find someone else who has a personal telecommunications system and ask for help. A problem can often be isolated by substituting a component that is known to be working. Problems that cannot be eliminated in this way usually turn out to be due to hardware or software incompatibility. It pays to call the technical support staff of the manufacturers of the equipment and software being used. Often, they have solved similar problems for others using the same combination of equipment.

BEYOND THE BASICS

The basics of telecommunications have now been presented. In Chapters 3 and 4 we present ideas and descriptions of many ways to use commercial services and information utilities in the classroom. Ideas on how to integrate telecommunications into specific curriculum areas are given in Chapters 5 and 6. In Chapter 7 we discuss how to generate support and develop management guidelines for building telecommunications into the mainstream curriculum. Additional technical aspects of telecommunications are presented in Chapter 8, and Chapter 9 looks toward the future and previews technical advances and educational practices that are on the horizon.

3
Simulations and Local Services

The use of telecommunications in schools evokes concerns about cost and ease of use. It is important to consider both of these factors when learning and using telecommunications. People often begin with large and expensive commercial systems because they are well publicized and easily accessible. Due to the cost and size of these systems, however, they are not always appropriate for beginners. A much gentler approach is to orient students with simulation programs and begin actual online activities with small, inexpensive, local bulletin board systems. This chapter explores such alternatives. It addresses three areas: the use of simulation programs for teaching telecommunications, the use of local bulletin board systems, and setting up a local bulletin board.

SIMULATIONS AND ELECTRONIC MAIL PROGRAMS

Programs that simulate online bulletin board systems provide opportunities for students to experience telecommunications before they actually get online. Because simulations can be used without phone lines or modems, they are a cost efficient way to introduce students to telecommunications. Many of the programs can be legally copied for use in a computer lab setting. Students who are new to telecommunications often feel a great deal of anxiety when online for the first time, particularly when costs are involved. Simulation programs help to reduce that anxiety by providing an opportunity to interact with a bulletin board system without time constraints or connect charges.

Examples of Telecommunications Simulations and Tutorials

Simulations are used successfully to train teachers as well as students. They are often used in workshop settings to familiarize teachers with telecommunications and in classrooms and computer labs with individuals and groups of students. Some programs are commercial, some are in the public domain, and still others, such as those from Softswap, are copyrighted but allow copying for educational use. The following programs are among those available. The Resource Section has additional information for each program listed below.

Electronic Village, published by Exsym, includes a simulation of a 300-baud bulletin board system. Students actually type in a phone number and password to log on to Gen-Com-Net. They are not able to leave mail for other users, but they do have a chance to read electronic mail, browse through bulletins on a variety of topics, and explore several other menu options. "Help cards" are available with all of the necessary commands as well as some useful definitions.

The Information Connection, published by Grolier Electronic Publishing, is another simulation program that offers users a very different experience. After familiarizing students with the concept and vocabulary of telecommunications, they are given four research questions and shown how to find the appropriate information from *Infoserve,* a fictional database.

SimuComm, from Softswap, is used to train students to do online data retrieval. It is a copyable program developed by Jack Gittinger at the University of New Mexico. It also includes an electronic mail feature and a bulletin board section.

ERIC MICROsearch is a program for high school and college students that provides an opportunity to practice data retrieval skills with the ERIC bibliographic data base. It is published by Information Resources Publications and includes data disks that are updated quarterly.

Window on Telecommunications, published by Exsym, is a tutorial that allows students to choose topics from a menu and see explanations of telecommunications terminology with graphics and examples.

Electronic Mail Programs

Another type of training involves an electronic mail system that needs only one computer and no phone line or modem. *Electronic Mailbag* provides up to one hundred students with the chance to write and receive electronic mail. An individual can log on and read mail, send mail, or get a user list. A simple editing system is available. Three different levels allow teachers to use the program with elementary through adult students. A teacher's guide that accompanies the program is filled with activities and curriculum ideas. Many options, including form letters and printing out user lists with passwords, are available to the teacher, who acts as the "sysop"—the systems operator, manager of the telecommunications hardware and software. More information about the role of the sysop and starting a bulletin board is given further on in this chapter.

KidMail, also from Softswap, is an electronic mail program that allows students to log on and send and retrieve mail. Like many public domain programs, the documentation comes on a disk and can be printed out. A unique feature of KidMail is that its developer, Wayne Ayers, has created a network of users. Classes that want to communicate with other classes around the country collect introductory messages on a KidMail disk. They can then contact Wayne and send their KidMail disk to him. He assigns a "pen pal" class and sends the disk to the pen pals so that they can log on and respond.

Modemless CMS (Computer Mail System) is another public domain electronic mail program that includes mail, message boards, and poll-taking functions. It resembles bulletin boards that run on CMS software and is helpful to teachers who are planning to use CMS software to set up their own bulletin board systems.

THE TRANSITION TO ACTUAL ONLINE EXPERIENCES

After students have participated in offline activities, simulations, and tutorials, they are ready for the transition to actual online experiences. Students should first become familiar with the telecommunications software they are going to use. The teacher should demonstrate the log-on sequence and any other information students need to get started. Most teachers choose to teach the more advanced functions, such as capturing files, after the children are comfortable with being online.

Choosing a Bulletin Board System

Thousands of bulletin board systems exist throughout the country. Their low costs and small size make them a good choice for beginning telecommunicators. Bulletin board systems accessed by a local phone call are usually free or have very low charges. Sometimes there is a one-time fee or a yearly membership. An example of this is the $10 charge for LES-COM-net, mentioned in Chapter 2.

The user environment of small bulletin board systems can provide a non-threatening experience for the beginning telecommunicator. When the group of users is from the same geographical area, it is easy for a caller to feel a part of the online community. Topics such as local politics, social issues, and economics are often discussed. When students and teachers are involved, the topics usually relate to educational issues and school activities such as science fairs, sports, and class schedules. Chapters 5 and 6 give several examples of how teachers use local bulletin boards in the classroom.

Often a small section can be set up on a larger system for a group of students from a specific classroom or school district. A password may be required to keep the section exclusive. One of the advantages to having a small section of a larger system is that a teacher can have control over a portion of the system without the responsibility of maintaining the hardware and software. Although a variety of systems are available in every city and offer a multitude of functions, teachers should be selective. If a bulletin board has many users and some of them are not people with whom students should interact, then it may not be a good choice, no matter how private the school's section is.

AVAILABLE FUNCTIONS ON LOCAL BULLETIN BOARDS

Electronic Mail

Electronic mail involves the sending of a private message from one user to another. Most systems provide a store-and-forward type of service, where the sender's message is sent and stored on the system until the recipient logs on and picks it up. The requirements for communicating in this fashion are very different from other forms of communication.

Messages should be kept short.

More than a screen full of text may scroll up the screen before the user can finish reading it. When an "autoreply" function is available, a short message can be seen while a reply is written. Example:

To: HILARY STONE
From: JOHN KING
Date: 7/2/89
Time: 7:23 a.m.
Subject: Money
I received the bill for the book fair and will send it to you through interoffice mail. Please let me know when you receive it and if it is correct. Thanks.
(R)eread (A)utoreply (N)ext Message

Messages should be unified in content.

When replying to an electronic mail message, it is easier to reply to one topic at a time. If many topics are going to be discussed, they should be sent in several different messages, or the topics or questions should be numbered so that the recipient is sure to respond to all of them. Example:

To: MARGE DALY
From: MARY WHITE
Date: 3/25/89
Time: 3:19 p.m.
Subject: Need information! Thanks for the info. about the Apple IIe purchase. I still need to know:
1. Were you able to get the mouse?
2. Did the Stickybear software come in yet?
3. Do you have enough blank disks?
(R)eread (A)utoreply (N)ext Message

Questions should be placed at the end of the message.

There is a better chance of it getting an answer if the question falls at the end of the mail message. Also, if a mail message ends with a question, there is a better chance of developing and continuing a dialogue. Example:

To: FRANK JONES
From: VIRGINIA GREEN
Date: 5/30/89
Time: 2:06 p.m.
Subject: Bear-ly enough
The field trip to the zoo went well, even though we had to leave early because the bears got loose. How was your trip to the aquarium?
(R)eread (A)utoreply (N)ext Message

Reference should be made to the previous message.

Details such as dates, times, and locations should be repeated in a response. Consider the following message:

To: SARA MOORE
From: JIM HARRIS
Date: 6/3/89
Time: 5:03 p.m.
Subject: Meeting
Sure, it would be great to have that meeting with you then. I will be out of town for the next couple of days and will see you Tuesday where we met last time. Don't forget to bring the papers I mentioned in the last message.
(R)eread (A)utoreply (N)ext Message

The problem with this message is that Sara receives about 15 mail messages each day and does not always keep a hard copy of them. She would not know from this message where and when they are meeting or what papers to bring. She may not be able to reach Jim before Tuesday to confirm the logistics. A better message would have been:

To: SARA MOORE
From: JIM HARRIS
Date: 6/3/89
Time: 5:03 p.m.
Subject: Budgeting Time
Sure, it would be great to have that budget meeting with you on Tuesday, June 7th, at 3 o'clock, in my office. I will be out of town for the next couple of days, but I will check my Email on Tuesday morning. Don't forget to bring last year's budget papers. Could you please confirm by Email? Thanks.
(R)eread (A)utoreply (N)ext Message

Sara does not have to refer to the previous message Jim sent, and has all the information she needs. She also is able to confirm the information with Jim before the meeting.

Care should be taken to avoid misinterpretation.

In face-to-face communications, people use a variety of methods to express themselves. In addition to words, they use facial expressions, body language, intonations, and sound effects to communicate their thoughts. Regional accents and physical appearance also have an effect on the listener. However, in telecommunications, people must depend entirely on the written word to convey their thoughts, so it is essential that messages be expressed as accurately as possible. A variety of techniques can be used. Important words or phrases can be typed in capital letters to add emphasis. (Remember, though, that when an entire message is written in capital letters it comes across as "yelling," so capitals should be used sparingly.) Sarcasm or joking used in telecommunications can often be misinterpreted, and feelings can be hurt. Children should be taught how to express their feelings so that they will not be misinterpreted. Actually typing "not really" or "just kidding" can help. Example:

To: MICHAEL ROGERS
From: BARBARA HALL
Date: 5/2/89
Time: 4:32 p.m.
Subject: Hot Air
I received your message about planning the end of the year party for Mr. Sargeant. Were you joking when you suggested renting a hot air balloon and having it land on the football field? I always did think you were a bit flighty. (Just kidding--I think it's a great idea.) When can we meet to plan the big event?
(R)eread (A)utoreply (N)ext Message

When telecommunicators type their messages while online, they often use abbreviations and shortcuts. Often misspellings occur and grammar is not always correct. Teachers may want to address this issue directly and discuss it with the students. One solution is to prepare messages offline using a word processor and edit them before sending them to the bulletin board.

There are times when abbreviations are used to express feelings or relay messages without a great deal of typing. This symbol :> can be turned sideways to make a happy face or turned the other way to make a sad face. "BRN" means "be right back" and "BTW" means "by the way." Other abbreviations such as "LOL" for "laughing out loud" and "ROFL" for "rolling on the floor laughing" are often used in a real-time conference.

Many telecommunicators create their own logo out of letters and symbols, save it on a disk, and place it at the bottom of their mail messages and bulletin board entries. Special occasions also prompt users to create text graphics such as the design shown in Figure 3.1.

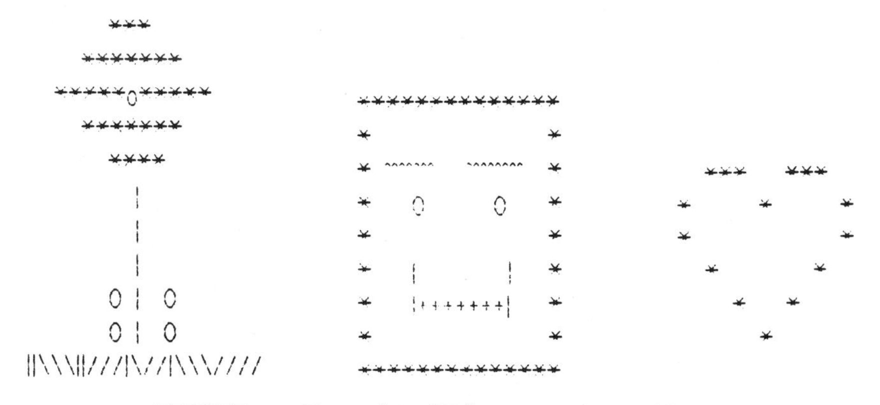

FIGURE 3.1. Examples of Telecommunicators' Logos

Special Features of Electronic Mail

Electronic mail systems often offer other functions in addition to sending and receiving mail. These depend on the bulletin board software that is run on the host computer and sometimes the creativity and programming skills of the system operator. The online editing that is available varies widely, but is often in the form of a line editor which allows the user to edit only one line at a time. Usually lines can be added or deleted, but only one at a time. If many errors are made, it is often more efficient to retype the message. The lack of sophisticated editing capabilities is another reason why messages should be kept short. Preparing messages offline using a word processor, saving them on a disk, then sending them once online is even more efficient.

Below is an example of options available on Boston's CitiNet online editor, for use after a message is created:

ADD	enter text
CHANGE	modify the current line
DELETE	delete a line
FIND xxx	find first (or next) xxx
FLUSH	clear the buffer without exit
LOOK	display text with line numbers
QUIT	cancel the whole operation
READ	display text without line numbers
END, EXIT	finish editing (or / or //)

Some systems have capabilities that allow a user to send mail to more than one person at a time, called "mass mail" or "junk mail." This can be very helpful for teachers who want to send mail to all of their students at once. Class assignments and school announcements can be sent this way. Also, a select group of students can be sent a message, such as an overdue library book notice. (It is a good idea to advise students that if they abuse this capability by sending mail that harasses people, their privileges will be eliminated by the system operator.) Following is an example of the Mass Mailing Utility Menu on Boston CitiNet. User responses are shown in boldface.

Mass Mailing Utility
Option Description

---

1 Enter Subject of Mailing
2 Enter Your Message
3 Enter Mailing List
4 Run Mass Mailing Program
Option (or Back, Top, MSG,?, or Exit):**1**

The Subject of Your Mail Message
Subject? **Party**

Enter Option or RETURN for menu: **2**
Enter The Text of Your Mail Message

Type HELP for help, EXIT to finish.
Type END to review and edit.

1> **November 2nd, 1989, 4:00 - 6:00 P.M.**
2> **there will be a party for all of the**
3> **volunteers who generously helped out**
4> **during the book sale.**
5> **It will be in the cafeteria and there will**
6> **be refreshments and awards.**
7> **Hope to see you there!**
8> **Mrs. Green, Principal**

9> **EXIT**

8 lines saved.

Enter Option or RETURN for menu: **3**
Enter Intended Recipients, (1/line)
Type HELP for help, EXIT to finish.
Type END to review and edit.

Type HELP for help, EXIT to finish.
Type END to review and edit.

1> **MEGKELLY**
2> **TOMBERNARDO**
3> **JASONWHITE**
4> **CONNIEDALE**
5> **VIVIANJONES**
6> **MARIONCOULTER**
7> **FRANCASTRO**
8> **EXIT**

7 lines saved.

Enter Option or RETURN for menu: **4**
Will send the following:
From: MRS. GREEN - Principal
Subj: Party
File: ypo:mm_Mrs. Green.txt (8 lines)
Sending to YPO:MM_Mrs. Green.LST . . .

1. To: MEGKELLY Posting MAIL to MEGKELLY
2. To: TOMBERNARDO Posting MAIL to TOMBERNARDO
3. To: JASONWHITE Posting MAIL to JASONWHITE
4. To: CONNIEDALE Posting MAIL to CONNIEDALE
5. To: VIVIANJONES Posting MAIL to VIVIANJONES
6. To: MARIONCOULTER Posting MAIL to MARIONCOULTER
7. To: FRANCASTRO Posting MAIL to FRANCASTRO

7 letters sent.

The list of recipients can be prepared in a file ahead of time and can be used over and over. It is wise to add yourself to the list as a way of confirming that the message was sent accurately.

Some systems have a "return receipt" function. When the recipient has read the message, another message is placed in the sender's mailbox telling him or her that the recipient has received the message and what time it was received.

To: JIM HARRIS
Date: 6/3/89
Time: 7:08 P.M.
-------Return Receipt-------

Message to SARA MOORE received.

Another helpful feature is the ability to get a directory of the electronic mail in your mailbox. An example:

Sender	Time and Date of Message	Subject
1 CATHY: Catherine Crane	2:22 PM 6/08/89	Software
2 BARBARA: Barbara Reese	1:16 PM 6/09/89	Help
3 GENE: Gene Nader	9:20 AM 9/09/89	Thank you!
5 FRANK: Frank Morris	3:33 PM 9/09/89	!

Some systems allow users to delete selected pieces of mail and store the rest. Others only allow users to save all of the mail or to delete all of the mail.

Electronic mail offers a wide range of possible uses. When choosing a bulletin board, make sure that it has at least the basic functions needed.

Bulletin Board Sections and Special Interest Groups

Traditionally, a special interest group is dedicated to a particular topic and is of interest to a small subset of the general user group. Visitors to the section post messages that are read by anyone visiting that section. They may be requesting information about telecommunications software or a particular type of computer.

Sections of a bulletin board can be reserved for the school lunch menus, schedules of classes, special school events, sporting events, and teachers' meetings. These are posted for anyone who has access to that area of the system.

Electronic information can be changed or updated by the system operator and be made available to the users immediately. The sysop can draw attention to an important piece of information or update by displaying a headline as part of the introductory screen, similar to an "Extra" in the newspaper world. Directions to the section where the detailed information can be found are usually included.

Forums and Discussion Groups

Many bulletin board systems include sections designed for the exchange of ideas and opinions. These provide teachers with a valuable opportunity to express themselves about important issues and discuss their ideas with others who may agree or disagree. A typical forum might be dedicated to one topic, such as a particular school discipline policy or a change in the school calendar.

Bill Niquette, a high school student and user of the Champlain Valley Union High School Bulletin Board, talks about one of the forums on the system:

> My personal favorite . . . is the Artificial Intelligence Forum. I have always been fascinated by the thought of intelligent computers, but never had known anything about the research being done, or the ideas behind it all. I began to read the messages left there, and before I knew it I was getting answers to all my questions, gaining all sorts of interesting knowledge directly from other people. Somehow, getting direct replies to your questions is much nicer than reading them from a book. Even nicer, is that, unlike a book, if you don't understand the explanation you get, it is easy to get someone to explain things better to you. (Niquette, n.d.)

Forums can become quite heated and sometimes a sysop needs to monitor the content and language of a particularly controversial topic.

Chatting

Some bulletin board systems have a chat mode, sometimes called "CB" as in Citizen's Band. In this mode, people can "talk" to other users directly in "real time," similar to having a conference call on the telephone. This is only possible when the system has multiple lines and can handle several people calling in at once. When "talking" to another person or to many people, confusion can result if there are not some conventions. Hitting RETURN on some systems leaves a space, signaling the others that a message is completed. On other systems, when one person finishes they type "stop" or "over" to tell the others that they are finished with their message.

The following is an informal chat with four people who meet on Chat frequently. They often make arrangements for meeting at a specific time and usually send out "invitations" to classmates.

/GREGORY: Let's start planning the pep rally!
/BOBBY: Last month's was great! Want to have it outside the gym again?
/JUDY: Sounds good 2 me.
/CHARLOTTE: Me 2.
/JUDY: How many people will be there?
/GREGORY: About 200.
/BOBBY: Judy, are you bringing the food?
/JUDY: For 200 people?! LOL
/CHARLOTTE: ROFL
/BOBBY: Who needs food when we have team spirit.
/GREGORY: With a 12-2 record our spirits are pretty low.
/CHARLOTTE: Sob, sob, our team needs a boost.
/GREGORY: Rah, rah.
/JUDY: OK so let's get together tomorrow PM, in the gym, to plan the rally.
/BOBBY: Fine - see you there after school.
/CHARLOTTE: OK by me.
/GREGORY: C U then.
/JUDY: Bye 4 now.
/BOBBY: Signing off.

Chatting is especially effective in a classroom setting when a discussion or meeting takes place with students or teachers in remote locations. The travel time and logistics of gathering people together for a meeting is difficult. Teachers from different districts can get together this way to share ideas and discuss issues. The interactions can be captured on a disk, edited in a word processor, and sent to the participants as "minutes" of the session.

One use of chatting in a classroom setting is to arrange for a guest speaker. It can be someone the class knows, such as the principal of the school, a community personality, or a representative from a particular profession. A historical figure such as Abraham Lincoln can be the subject of a chat. Sometimes the person on the other end of the line is a mystery figure and the students type in questions to gather information in order to guess the person's identity. The following chat was in the form of a twenty-questions game, with the children asking "yes" or "no" questions. The only information the children received was that the mystery person was someone they knew.

/BRUCE: Do you work in our school?
/MYSTERY PERSON: Yes.
/JIMMY: Do you work in a classroom?
/MYSTERY PERSON: No.
/KEITH: Do you see the fifth grade every Thursday?
/MYSTERY PERSON: No.
/PAT: Do you wear a uniform?
/MYSTERY PERSON: Yes.
/ANN: Is it white?
/MYSTERY PERSON: Yes.
/MICHAEL: Do you check people's ears?
/MYSTERY PERSON: Yes!
/SUSAN: Are you Mrs. Letterman, the nurse?
/MYSTERY PERSON: Yes!

Often, a discussion about formulating questions evolves from this activity. Questions that elicit information to narrow down the possibilities are the best, and children should be encouraged to gather information before they guess at the answer. The children can also play the "mystery guest," either being themselves, a fictional character, or a well-known celebrity.

It is not necessary to use a bulletin board system to have a chat. Some telecommunications software supports remote operation and allows two computers to "talk" to each other. One computer can be in the classroom with students taking turns typing messages and questions, and the other can be in another location. The online interaction can still be captured for later retrieval. Chapter 5 gives examples of how teachers use this feature.

Additional Features

A list of recent callers and the times they called is available on some systems. This is useful when teachers want to find out which students have and have not logged on. They can use the information to monitor the use of the system as well as the volume of use at particular times. Another option on many systems is the ability to print out a list of all the users of a system. This is particularly useful for classroom settings where the user list can be posted next to the computer for easy access. On some systems the sysop can be paged. If the sysop is in the room with the host computer and wants to talk to the caller, then he or she gets online and types messages interactively with the caller.

Getting Acquainted with a Bulletin Board

One method of familiarizing new users with a local or commercial bulletin board system is to have an online scavenger hunt or trivia contest. With a multi-line bulletin board system, several people can go online at one time, which makes real time competition possible. Questions are prepared ahead and should be from a variety of areas of the bulletin board. Teams of four or five can work together to develop a strategy for researching answers. Sample questions, similar in format to the real questions, can be given to the teams so that they can go through a "dry run." It is important to formulate questions so that answers are not ambiguous or confusing.

One method of conducting the hunt is to have each member of the class in a different location. Teams divide up the questions and work independently searching for the answers. They communicate online, with one team captain collecting the answers and sending them by electronic mail to the teacher. The winning team can be determined by the accuracy of the answers and the time it takes each team to respond. If teams work together at one computer, each member might be an expert of a particular area or topic.

Boston CitiNet offers a range of forums and special interest groups. The following questions were used during a Scavenger Hunt on Boston CitiNet and reflect the variety of information available on the system:

1. What exhibit is currently at the Museum of Fine Arts?
2. Who is the editor of the poetry section on CitiNet?
3. Who on CitiNet is trying to sell a parrot?
4. What is the number for the Consumer Protection office?
5. According to the CitiNet weather forecast for tomorrow, what will the temperature range be?

6. What is the annual percentage rate of Bay Bank for a 15-year fixed mortgage?
7. What college is presently offering a course in telecommunications?
8. What rock group is playing at The Rathskeller this weekend?

Participants were successful gathering the information and relaying it to the team captain via electronic mail.

If a bulletin board system has only one phone line, a hunt can still be done, but in staggered time slots. Each team gathers at one computer and team members each have specific roles. The teacher places the questions in the team captain's electronic mailbox at a specified time, and the team has a designated amount of time to hunt for answers. The team captain then reports the answers by electronic mail to the teacher.

Much excitement is created by this type of activity. Participants learn the various sections of a bulletin board very well and quickly become skilled in moving around a system.

UPLOADING AND DOWNLOADING

Many bulletin boards have download sections where users can have a program sent from the host computer to their own computer to save on their own disk.

Public Domain Programs

Thousands of free, non-copyrighted programs are available. Public domain programs include games, utilities, and educational software. They vary widely in quality and usefulness. One very popular series of public domain software is FrEdWare, Free Educational Ware, including FrEd-Writer, FrEdSender and FrEdFiler. These programs allow schools with limited budgets to acquire application software that can be copied and used in a lab setting and distributed to teachers and students for use at home as well.

Shareware

The writer of a shareware program distributes the program to bulletin boards and user groups. Although the program may be easily obtained, the writer expects the user of the program to send in a payment. The average price for a shareware program is much less than a commercial program. A shareware application, such as a word processor or telecommunications program, often costs under forty dollars.

Red Ryder is a shareware telecommunications package that has undergone numerous revisions and has gained much respect. It is compared favorably with programs costing many times its price. Documentation is available by printing out information from a disk and technical support is available by telephone. A caller often gets the author of the program and has immediate access to helpful advice and information.

Drawbacks of public domain and shareware software have been the lack of written documentation, the absence of technical support, and programs that are incomplete or not carefully checked for errors. A more serious issue is the danger of viruses. John Dvorak described the problem in *PC Magazine:*

> A computer virus (sometimes called a Trojan horse or a worm) is a small and sinister piece of software code that literally infects your machine. It is inserted into a public domain or bootleg program and, when the program is used, the virus is alerted and rewrites itself into something in your system and typically (and eventually) calls a hard disk routine and tells the disk to erase itself. Computer sabotage. It's getting worse and we should all be aware of it. (Dvorak, 1988)

Kristi Coale, Assistant Editor of *MacUser*, describes them further:

> Computer viruses, not unlike the biological variety, are self-replicating. They have a self-replicating code and get into resources such as INITs and CODE. So far, we know that they spread from one computer to another (and within this, from System file to System file) through infected applications. You can get infected applications from online services, from bulletin boards, and from disks you swap with friends. Once in a new host (your Mac), the virus spreads to other places, usually System files, where it may lie dormant for days, weeks, months, or even years before wreaking its havoc. (Coale, 1988, p. 305)

Macintosh computers are particularly susceptible to viruses. "Scores" is a common, unpredictable virus that causes system crashes, printing problems, and the modification of certain programs. Another less common virus, "nVER," also modifies application programs.

Numerous public domain and commercial programs are being developed to detect viruses. Detection of such viruses can be accomplished with a public domain program called Interferon 2.0, written by Robert Woodhead. Another such program is Ferret 1.1, free through CompuServe. Others are found on GEnie and Delphi online information systems (see Chapter 4). However, as Thornburg reminds us:

> Before you get your hopes up, however, it's important to realize that the typical vaccine is able to prevent only certain viruses and that computer hackers are constantly developing new viruses that will outsmart the vaccine programs that are already in existence. (Thornburg, 1989, p. 18)

Removal of viruses can be accomplished by the removal of any infected applications and then reinstating the software with a clean copy.

Reinitialization of a hard disk is recommended. Virus prevention can come in many forms. Use original, rather than copied, versions of software. Check newly attained software, especially public domain and shareware programs, before using them. Lock application programs so that viruses cannot affect them.

The Piracy Issue

Unfortunately, illegally copied commercial programs are available on many systems. Usually the sysop of such a bulletin board system does not make these obvious or available to the new user, but only to those users who have proven capable of providing even more valuable programs for distribution. Breaking the copy protection on a commercial disk is a skill that is valued in communities of pirates, and individuals who have proven themselves skilled in this area are given upgraded privileges on the system and access to the valuable and hard-to-copy software.

Often the name of a bulletin board system, such as Pirate's Passion and Captain Hook's Cove, is an indication of whether it contains illegally copied software. Some bulletin boards appear to have only public domain software but also have a hidden section with illegal programs. When a user demonstrates interest and has shown that he or she has other programs to offer, the user is given upgraded privileges and allowed into the sections containing the more valuable programs. Stories such as the following are heard often:

> When I first logged on to a pirate board I had a great time downloading game programs and giving them to my friends. I had heard that they had more programs than were on the main menu and asked someone on the board about them. I was asked what programs I had to offer. I had nothing that the board did not already have and told them so. That night when I tried to log on I found that my password did not work.

Many young users struggle with the illegal nature of illegally copied software.

> It was really fun to collect disks full of programs from the bulletin boards I logged on to. I had hundreds of games and many educational programs. My parents couldn't afford to buy me these programs and I thought that because I was learning something from them that it was OK. I didn't think I was hurting anybody and didn't think I was doing anything wrong. When my computer teacher began to talk about pirated software I realized what I was doing. She discussed with us how copying commercial software was like stealing and how many small companies have gone out of business because they sold fewer

copies of their products. I know I shouldn't use these bulletin boards any more and am trying to stop.

It is difficult for children to stop copying valuable software and to fight the peer pressure that supports their "habit." Computer specialists and other school staff can play an important part in educating children about the illegal nature and consequences of copying commercial software.

The issue of piracy is dealt with in a software package called *Ethics Online*, published by Exsym. A slide show presentation on disk provides material for discussion. Cards describing situations are used for role playing. Among the situations presented are the illegal copying of a commercial program, using someone else's password without their knowledge, spreading rumors on a bulletin board system, and breaking into a school's mainframe computer to make a grade change. The situations are described and the participants are asked to take on the roles of both the student committing the act and the student trying to reason with the "criminal." Often acting out the situation brings up possible solutions and can be helpful to students and teachers when a similar situation occurs in real life. The standards suggested by *Ethics Online* include:

- Avoid the pirating of commercial software.
- Avoid the use of profanity.
- Avoid spreading rumors and false information that can be harmful to others.
- Honor the right to privacy.
  Do not read the private electronic mail of others.
  Do not use another person's password or I.D. number.
- Do not attempt to crash or harmfully penetrate any computer system.
- Do not break any laws with what we say or do online.

Teachers should encourage students to discuss these guidelines and the moral and ethical issues that they reflect.

EXAMPLES OF EDUCATIONALLY ORIENTED BULLETIN BOARD SYSTEMS

Systems around the country vary widely in their content, objectives, and organizational structure. The system operators, or sysops, might be teachers, administrators, students, consortium directors, or a team consisting of some of these. Some systems include the parents of students in the school and other citizens of the community who have access to a computer and a modem. The following are some examples from the many local systems.

A Student-Managed Bulletin Board System

The Champlain Valley Union High School Electronic Bulletin Board serves school systems in and around Essex, Vermont. It was organized by the computer club advisor, Craig Lyndes, and the members of the computer club. In September 1986, IBM awarded the computer club a grant of $3000 that enabled them to purchase their initial equipment. Ongoing fund raising activities are needed for additional funds. The board serves over 400 users and has more than 30 different special interest groups and forums. Three student system operators take turns maintaining the system and meet to deal with the issues of access, improvements, and online behavior. A constitution was written up by the students and their advisor that provides guidelines and legislates enforcement. The board is used for communication among the schools in the district as well as between schools and children's homes (Lyndes, 1988).

A Consortium-Managed Bulletin Board System

Many bulletin board systems are run by district and state teacher groups. MeLink is run by the Maine Consortium and provides services to teachers throughout the state. It is run on a personal computer from the Auburn, Maine, office and provides teachers from small, remote towns opportunities for communication and the sharing of ideas. Teachers around the state recently used the service to collaborate on a grant proposal. Without it, the teachers could not have accomplished the task in the allotted time.

A Teacher-Managed Bulletin Board System

Fred Wheeler, Behavior Management Specialist in Norristown, Pennsylvania, began a bulletin board for his socially and emotionally disturbed senior high school students which now serves eleven schools. Wheeler describes the success of the system:

> Since most of my students have academic as well as emotional problems, the goal of the project was to motivate them to write more frequently as a means of increasing their language art skills. The daily assignment in telecommunications was to access the BBS, look for mail and then to send a message of at least three sentences. Sounds easy, simple, and fast but the students agonized over who to write to and what to say, often taking fifteen minutes or more.
>
> A student who disliked writing anything, grudgingly used the computer to write some assignments. When I noticed that he could hardly wait to check his mail on the BBS, I arranged with another teacher to have other students write and ask him to reply! It worked and he was hooked! (Wheeler, n.d.)

For many students, the bulletin board provides a method of communicating with classmates, even after traditional classroom situations are altered.

> Shortly after the school year began, one student was transferred to another classroom. For an emotionally disturbed child this represents a major trauma. However, knowing that she could use the bulletin board to write to her teacher and friends in her old class eased her transition. Gradually her need for this link lessened and she is now fully integrated into her new class where she is considered an expert in using the bulletin board. (Wheeler, n.d.)

A Bulletin Board with Educational Possibilities

Some bulletin board systems are not designed with education in mind, but can provide opportunities for valuable educational experiences. SeniorNet, part of the Computers For Kids Over Sixty Project in San Francisco involves senior citizens, ages 60–95, communicating about numerous issues including those of special interest to older adults. Intergenerational activities have been organized where young people communicate with older ones, which helps eliminate many of the traditional stereotypes.

A Multi-Use System with a Section for Educational Purposes

Local boards with multiple lines and many uses sometimes have a section dedicated to education. Boston CitiNet has thirty phone lines and is supported by advertisers, making it possible to offer the system at no charge. A yearly mailbox usage fee is discounted to educators. Teachers are encouraged to use the system for educational projects and are supported with privileges such as mass mailing. Numerous educational activities have been fostered by this system and have included a debating project between students in the towns of Sharon and Canton, Massachusetts. Half of each team consisted of students from Sharon with the other half from Canton. This division necessitated communication among the children while they were researching their debate topics and planning their presentations. Files were sent back and forth, edited by both halves of the team. Electronic mail was also used. The culminating activity was an actual debate in one of the school's auditoriums, followed by a reception where the students showed their parents how to log on to CitiNet.

HELPING STUDENTS TO LOG ON TO A BULLETIN BOARD

Students should be instructed in the basics of telecommunications (a tutorial is given in Chapter 2) before they try to log on to the system. First-time callers are usually asked if they are a first-time caller, followed by a series of questions such as: their full name, address, and telephone number; the type of computer and modem they are using; and their particular configuration. Callers are assigned a password and are given limited privileges. A sysop may check the telephone number to verify that the callers are telling the truth when they logged on and to ensure that callers are not impostors. In many systems there is a twenty-four hour delay before users are assigned full privileges on the system and can access all choices on the bulletin board's menu.

After a user has received full privileges, she or he can log on without the initial interrogation. After the welcoming screen the caller is asked to input her or his username and a password. Because of the issue of privacy, a password is not usually visible; either nothing, an "X", or "*" appears on the screen for each letter. The caller can then choose to read mail or to go to one of the areas of the system represented in the main menu. Most callers want to read their mail immediately, and some systems place the user in the electronic mail section of the board when they log on and have mail waiting.

The organization of bulletin boards and methods used for a caller to move around the system vary greatly. Many systems have a main menu with choices that are displayed on the screen and a prompt at the bottom. Users are requested to type in their choice. They are then given a submenu with additional choices.

It is important that students be shown how to log off properly. If a caller ends the online session by turning the computer off or unplugging the modem, they are often still logged on to the system and online charges continue to accrue. Some systems automatically log off a user if he or she has not typed any keys in a certain amount of time. Chapter 2 gives one "panic button" way to log off a bulletin board in case the prescribed method is not known.

New users should read through the help section of a board if one is available. Sometimes additional help menus are found in the various areas in the system. Often, the best way to learn about a new system is to communicate with other people on the system by leaving bulletins or sending messages to the sysop.

STARTING YOUR OWN BULLETIN BOARD SYSTEM

When local bulletin boards are not available or do not meet the needs of a particular school or school district, then starting a systemwide or schoolwide bulletin board may be the answer. Being a system operator can be an exciting adventure for a teacher, group of teachers, or in some cases a group of students. The freedom to design a bulletin board to fit the unique needs of a particular group can add to the usefulness of a system.

Chris Clark, system operator of The Electric Slate, in Norwich, New York, lists reasons why someone should set up their own BBS:

- Telecommunications is not "the wave of the future"; it is standard procedure in business now, and our students need to learn about it.
- A BBS is the least expensive way to give kids real experience with telecommunications.
- It can be used to promote student writing.
- It's a useful communications tool for students, teachers, and administrators alike, reducing paper flow and telephone tag.
- If the board is open to the community, it can be a good public relations tool.
- It's fun, so it's motivational. (Clark, 1988, pp. 11–12)

The responsibilities of the system operator are varied and time consuming. Usually, the more creative and motivated the system operator, the more time he or she spends on the system. It is an open-ended, and therefore unending, job. Too often bulletin boards are a short-lived phenomenon because of lack of interest or energy on the part of the sysop. The investment in equipment is substantial and the investment of time often seems unlimited. A dedicated sysop provides a service to all of the users of the system and can enjoy a rewarding experience if the users are satisfied with the services provided. On the other hand, many sysops have trouble dealing with the complaints and requests, not to mention the hardware and software problems that occur. Maintenance of a system can vary from a few minutes a day to endless work on updating and improving the system. The following is a common scenario:

> I work every night and most weekends on the board. The day to day tasks of welcoming new users, cleaning up files, and making backups become routine after a while. The real challenges come with my desire to keep the board lively with exciting information for students and teachers. I try to think up creative ways to involve people and motivate them to contribute to the board. Nothing is as rewarding as when students start to suggest new forum topics and then go on to support them and take responsibility for them. Several of the forums are now moderated by students.

In addition to the maintenance of a system, the sysop must deal with ethical issues of piracy and privacy. Even when uploading and downloading are not a function of a system, there are issues involved with the privacy of passwords, the use of profanity, and the harassment of individuals or groups. Some systems have a small group of people who are in charge of developing and enforcing policies designed to deal with such issues. In larger systems, there are many sysops, each taking charge of a special interest group or particular function of the system.

Teachers who run bulletin board systems need a great deal of time to work on them. The best situations occur when the school's administration understands the demands on a sysop and allocates time, financial support, and recognition. A class period, at the very least, built into a teacher's schedule for work on the system can help, as well as enlisting the aid of student and parent volunteers for tasks such as editing files or backing up disks. Following are comments a teacher/sysop who has not been given this type of support might make:

> With increasing job responsibilities, it is not as easy to find the time either during work hours or after to keep the BBS serviced. Another part of this problem is that I am not able to give the time needed to provide emotional support to those who use the BBS. Being a computer maven is nice and occasionally receiving tacit recognition as such is okay but over these six years I have watched as others, who began their computing when I did and who were also local gurus, were hired as computer specialists or computer coordinators. They have official recognition and compensation for their skills, knowledge, and service. I find that my enthusiasm is waning. Is it any wonder that teachers jump ship and look for monetary fulfillment in the business world?

On the other hand, teachers who are given support and recognition for running a BBS might express a very different attitude:

> Learning about computers opened up a whole new world to me. Telecommunications was the topping on the cake. After starting our bulletin board system, I had the satisfaction of seeing teachers and students interacting with each other in a way only the phenomenon of online interaction can evoke. My favorite forum is the one where teachers, students, and administrators discuss school policy on such timely issues as drinking and driving. The positive public relations that have resulted from the BBS have helped to raise money for the hard disk drive we just purchased. The local newspaper will soon be writing an article about us. I am pleased that other teachers have begun to use the board for classroom activities and are becoming more involved. My job description has changed and I am now being given four planning periods each week to work on the board. It has become a creative outlet for me and I am thrilled about being involved in such an exciting venture.

Hardware and Software Requirements

The minimum requirements for setting up a bulletin board are a computer, a phone line, a modem, and bulletin board software. Some systems run with an Apple II+® computer and a 300-baud modem. A monitor is not required all of the time. Some maintenance can only be done by the system operator at the host computer. Although a monitor may be needed at that time, it can usually be removed between maintenance operations. A clock card can be inserted so that messages can include the time they were sent.

Enhancements include more sophisticated software and hardware, additional memory, and multiple phone lines. As the number of users grow, so do the demands on a system. Often, during prime times, a phone line can be busy for hours. A system operator can usually limit the time a user spends on the system as well as the amount of times a user can log on in a single day.

Software to run bulletin board systems varies widely. Programs that allow teachers to run a bulletin board should be easy to set up and maintain. A system of backing up the mail and bulletins is essential and should be part of the routine maintenance. Most software allows customization and permits the system operator to design original menu structures, opening screens, and the organization of bulletin boards. Also, prompts and other messages to the users can often be modified by the system operator. The freedom to modify the program allows the system operator to create a friendly environment and gives a bulletin board a unique character.

Local bulletin boards and simulation programs can provide a rich environment in which teachers and students can learn about telecommunications, communicate with others close by and far away, and expand their educational and social horizons.

REFERENCES

Clark, C. (1988, June). Confessions of an educator/sysop. *The Computing Teacher*, pp. 11–12.

Coale, K. (1988, September). "Razor blades in Apples." *MacUser*, pp. 305–314.

Dvorak, J. (1988, February 29). Virus wars: A serious warning. *PC Magazine*, p. 71.

Lyndes, C. (1988, April). *Student operated bulletin boards*. Paper presented at the Distance Learning in the Northeast Conference.

Niquette, Bill, 183 North Street, Winooski, VT 05404.

Thornburg, D. (1989). A personal computer plague. *Classroom Computer Learning*, 9(5), 18.

Wheeler, Fred, Behavior Management Specialist, 1682 Sullivan Drive, Norristown, PA 19401.

4
Commercial Services: Online Databases and National Services

Imagine, instead of spending hours in library, being able to access a list of 250 articles relating to telecommunications in elementary schools in ten seconds. Imagine asking for information about an adaptive device for a child with motor problems and receiving fifteen answers from across the country overnight. Imagine playing a computer game with a fourth-grader a thousand miles away.

These scenarios need not be merely imagined but are, in fact, possible today and have been for several years. Commercial services with hundreds of users have been in place and have progressed in sophistication in the past decade. Services such as shopping at Bloomingdale's and ordering airline tickets are not fantasy—they are available now and are expanding.

In addition to commercial systems providing the above functions, there are large databases of information that can be used for classroom activities and general research. Commercial databases provide an opportunity to access a vast amount of information very quickly. The information is updated often and is therefore quite current. It is easy to access and can be saved for later use. Without proper training and preparation, however, the confusion and cost involved can be substantial.

Commercial database services differ from local electronic bulletin boards in a variety of ways. Commercial systems have a large, varied user base in all parts of the country. Database systems have enormous amounts of information that can be accessed in a variety of ways. As the name implies, there are costs involved that are significant enough to deter many school systems from implementing a commercial online service. Many functions, however, are similar to those found on smaller bulletin board systems. In addition to information access, systems offer electronic mail, special interest groups, and discussion groups. This chapter outlines the benefits of using online databases, criteria for choosing a particular service, and an overview of classroom use and services appropriate for educational purposes.

BENEFITS OF COMMERCIAL DATABASE SERVICES

Commercial online databases contain thousands of volumes of information on every topic that a traditional reference library contains. No school library could possibly have the amount of reference material that is available on a commercial database. Traditional printed research materials are only as current as their date of publication. Online information can be changed quickly and is available for the user to access immediately. Although the ease of use varies widely among database systems, the information can be accessed more quickly than with traditional research materials. Peter Cook, who has worked on both the print and electronic versions of the Grolier Academic American Encyclopedia, compares the two:

> Clearly, a printed encyclopedia is limited because of the physical constraints of the print medium. The information on its pages remains fixed and cannot be dynamically rearranged to suit the needs and convenience of the end user. . . . But printed encyclopedias have yet another limitation that is particularly relevant when one considers the encyclopedia's primary role as a reference source, and that is the inaccessibility of much of the information. For example, to find information on Mark Twain in the AAE, a user could either access the T volume directly, and turn to the article "Twain, Mark," or use the index volume. The AAE index lists six references to articles on Twain—the one principal article and five additional articles that have been selected by the indexer. In fact, a search of the electronic edition of the AAE reveals references to some 25 articles that are of relevance to Twain. The printed index is limited by size and the selection criteria of the indexer, and cannot direct the user to all the relevant information.
>
> However, the printed encyclopedia has one advantage over the current electronic editions: it contains several thousand illustrations—photography, maps, paintings, diagrams, etc.—none of which can be stored and distributed economically using the electronic distribution media described so far. (Cook, 1987, p. 245)

The CD-ROM version of Grolier's Electronic Encyclopedia is widely used and an audiovisual encyclopedic database is being developed.

Although more expensive than smaller, local services, the national services offer a larger community of users and the advantage of having users from around the country with different perspectives. A teacher in Omaha can post a request for information about a particular piece of software and have her message read and responded to by a geographically diverse group of teachers. Chapter 5 gives examples of such an educational use.

CHOOSING AN ONLINE DATABASE SYSTEM

It is wise to be well-informed before making a decision about which commercial service to use. Most services have toll-free numbers and sales representatives to send information. Talking to other teachers who have used a system and reading reviews of the services in journals are other ways to gather information.

Cost Factors

There are a variety of financial considerations including initiation fees, hourly fees, and royalties.

- Initiation fees are on a one-time charge when a caller subscribes.
- Hourly fees are usually higher during prime-time business hours.
- Royalties are paid to certain authors when their material is accessed.
- Special offers for educators are often available to school systems including lower rates, multiple passwords, and curriculum materials for teaching about a system and developing strategies of online searching.

Types of Databases

Many systems allow the user to access specific databases or to let the service choose the appropriate databases. Specialized databases, such as LEXIS for the legal community, include information meant for a small population with a special interest. General databases, such as Bibliographic Retrieval Service (BRS) include a wide range of subjects and a variety of sources. See the Resource Section for more detailed information.

Formats Available

Services offer a variety of formats. Many searches result in a list of bibliographic references which include only the name and author of the article and little other information. Others include abstracts and a variety of other information. The following citations were found when searching the ERIC database through DIALOG for information about "Telecommunications and Elementary Education." The citations represent two of the many possible formats available.

5/3/3
EJ295101 EA517502
Closing the Gap.
American School and University, v56 n7 p41 Mar 1984

5/2/12
ED257595 RC015258
Primary Distance Education Population: Problems and Prospects. Research Series No. 2.
Taylor, Peter; Tomlinson, Derrick
National Centre for Research on Rural Education, Nedlands (Western Australia).
1984
206p.; A study of the pupil population served by primary distance education, the perceived needs of their home tutors and the adequacy of support services provided by distance primary schools in New South Wales, Queensland, Western Australia and the Northern Territory.
Sponsoring Agency: Australian Commonwealth Schools Commission, Canberra.
Report No.: ISBN-0-86422-051-4
EDRS Price - MF01/PC09 Plus Postage.
Language: English
Document Type: EVALUATIVE REPORT (142); TEST, QUESTIONNAIRE (160)
Geographic Source: Australia; Western Australia
Journal Announcement: RIEOCT85
Target Audience: Practitioners
Descriptors: *Access to Education; Classification; Correspondence Study; Delivery Systems; *Distance Education; Educational Needs; Educational Technology; *Elementary Education; *Enrollment Trends; Foreign Countries; Home Schooling; Mothers; *Parent Role; Parent School Relationship; Questionnaires; Rural Education; Rural Family; Services; Student Characteristics; Telecommunications; Tutoring; Tutors
Identifiers: *Australia; Australia (New South Wales); Australia (Northern Territory); Australia (Queensland); Australia (Western Australia); *Isolation (Geographic); School of the Air (Australia)

Some services produce a full text article. The text can be printed out while the user is online, which is the slowest method. It can also be captured and saved on a disk, then printed out with a word processor. With some services, the full text of an article can be printed out at the host computer and mailed to the user. This option is less expensive but can take several days.

Ease of Use

The menu structures and commands used can be logical, using simple English commands, or less obvious, involving control characters and numerous symbols. Searching can take place with menu-driven choices or long lines of code. Training is necessary for all systems, but some systems require only a few minutes to become familiar with them while others require extensive training. Research librarians may need thorough training to take advantage of a service such as DIALOG, but the benefits are great. Many of these services offer an easy-to-use version, such as DIALOG's Knowledge Index for new users and young students.

Support for Students and Teachers

Learning an online database system can be difficult for teachers as well as students. Support comes in a variety of forms and includes essential and helpful information. Most programs have written documentation. Correct settings for specific computers may be necessary and information such as the available baud rates, the word length, and the parity are essential. The phone numbers used to access the system are often listed in the written documentation or can be obtained through a local or toll-free number. The Telenet network, for instance, has both a written guide with all of the local access numbers, as well as an online list that is searchable by state or area code.

Once a user is online and presented with a choice of options, it is easy to become confused. One of the most frustrating problems is getting "lost" in a system, traveling to a place with no apparent escape. Having the available choices printed on the screen at an option prompt is most helpful. Online help can also be in the form of hitting a carriage return at a prompt to see detailed explanations of the possible commands. Many services have online tutorials and whole sections devoted to learning the command structure as well as shortcuts to using them. Some systems have various levels, with "express" modes for the experienced users. The advanced versions include fewer menus and allow users to travel around the system more rapidly. Simulation disks for specific systems are available sometimes.

Hands-on classroom training is available for the larger systems. In addition to workshops at conferences and inservice classes on database searching, there is training offered by the individual services. DIALOG offers courses in its extensive searching language at various locations several times a year. For an additional cost, the trainers will come to a site to train a group of users.

CLASSROOM USE

Although the ease of access varies widely among systems, the ability to master the skills necessary to search for information can be acquired by students from upper elementary school to college.

Minimizing the Costs

A major concern voiced by many teachers and administrators is the cost of online activities. The key component to saving online costs is keeping the time online to an absolute minimum. If users are well prepared and have their search strategies planned before they log on, the search has a better chance of being a success and the actual online time can usually be kept to a few minutes. Among the ways teachers can prepare children for online searching are with training in library skills and with simulations.

Using Electronic Databases in the Classroom

In almost every curriculum area, accessing online information can enhance a lesson or project. Listings of articles from magazines, newspapers, and journals can be used by students and teachers to augment knowledge and provide information otherwise not easily attainable. A student using telecommunications for the first time to gather information about a presidential election might describe her experience with an online database as:

> I got all my information ready, the key words I might need, the instructions on how to log on, and the description of what I needed to find out. My teacher showed us how to use the computer to search for information and did an online search in front of the class on the big monitor. When I got to go to the computer I was a little nervous that I might waste money or goof up. But it was real easy. I couldn't believe it! I got over eighty references the first time then narrowed my topic and got thirty-two. I saved them on a disk and logged off. The whole time online cost only three dollars. I went to the library and found enough of the articles to write my paper with many different viewpoints. I felt that the computer really helped me and I can't wait to work on it again.

Finding references to helpful information sometimes can be frustrating, however. A reference to the "perfect" article may be exciting to a child doing a project, but disappointing when the original document cannot be found. Although an abstract is often available, a printout of the full text may not be:

> I did everything my teacher taught us to do. I found out about my topic beforehand and prepared a bunch of key words. I worked on a search strategy with a lot of alternative plans. After waiting several days for my turn on the computer, I logged on and got over seventy bibliographic references. The trouble was, when I went to the school library to find the actual articles I couldn't find too many because the library didn't carry the magazines. I was able to find some of them at the big public library and do a pretty good report.

It is important for teachers to deal with these issues and help children decide when they should get entire articles. Sometimes abstracts are available and sufficient for students' needs.

Classroom management issues can lead to success or frustration. The availability of a phone line is one of the crucial ingredients to the success of an online research program. When a classroom has to share its phone line with for instance, the nurse's office, or when a switchboard causes delays, the wait for access can dampen the enthusiasm of any student. Although not all classrooms can have a direct phone line, there are solutions that can prevent conflicts. A schedule of usage can promote successful sharing. A long extension cord (but not longer than 100 feet) can be used to connect a computer in a classroom with an available phone line in another part of the school.

Should the computer and modem be used by children one at a time or demonstrated by the teacher in front of the class? Most experienced telecommunicating teachers agree that children learn best when they actually log on themselves. Many teachers demonstrate a log-on and search using a large monitor or "daisy-chained" monitors. The students then prepare their search and go online one at a time.

Roxanne Mendrinos, media specialist at Thurston Junior High School in Westwood, Massachusetts, has been using CompuServe for four years. She first trains students to efficiently search for information in a library using traditional reference materials such as the card catalog and the Reader's Guide. Much work is done with printed reference materials such as a thesaurus, almanac, and encyclopedia index. Students must be prepared with five key words and two pages of information about their topic before they log on. They must plan a search strategy before they attempt to actually look for the information and be ready to make instant decisions once they are online.

Having mastered these skills, they are given practice with the Boolean operators and shown examples of online searches. The emphasis on offline planning is essential for reducing the online time and charges. When students are finally ready to go online, they have been well trained in the necessary techniques. They feel comfortable and end up finding the information they need quickly and efficiently. The budget for online searching at Thurston Middle School last year was $250 for CompuServe and $400 for DIALOG. All 250 seventh- and eighth-graders had online experience. Ms. Mendrinos also uses the CD-ROM version of the Grolier's Electronic Encyclopedia with her students. Similar search strategies are used, but the pressure of using online time is eliminated. Because of these and similar activities, school administrators are beginning to consider the possibility of using money spent on traditional encyclopedias to pay for online services and CD-ROMs (Mendrinos, 1988).

Written documentation, online tutorials, and simulation disks are often available to bridge the gap between traditional and online searching. Search terms, logical connectors, and choices involved are necessary information. Familiarity with the options of a particular service is essential.

Boolean Logic

Boolean logic and logical connectors are discussed in many mathematics classes but are seldom used in a practical setting. These concepts are essential when searching for information in most systems. The speed at which students can access information often depends on their expertise in this area. Offline exercises using "and," "or," and "not" to delineate subsets of a group can be done with paper and pencil or, better yet, with manipulative objects such as attribute blocks. Judah Schwartz states his concern about the teaching of these skills:

> Boolean connectives are seemingly not understood. In general the competent use of databases requires a careful, rather than sloppy understanding of the words "and," "or," and "not." I have watched youngsters not understand why a database on United States presidents, when queried about the number of presidents born in Massachusetts and Vermont, insisted on claiming that no presidents were born in Massachusetts and Vermont. Clearly, the problem has nothing to do with the technology. Rather we need to educate people to use the language with much greater precision than they are presently accustomed to using. (Schwartz, 1987, p. 70)

Consider, for example, the search described earlier in this chapter on "Telecommunications and Elementary Education." Because of the many versions of the word telecommunications such as telecommunicate, telecommunication, and telecommunicators, the word was truncated in order to locate references to all of them. Searching terminology often includes asterisks as a type of "wild card" so that a search will include all forms of the word. Hence, the key words TELECOMMUNI* and EDUCAT* were used to find the citations.

Students who have library and dictionary skills will be better equipped to do online searching. Activities involving key words, logical connectors, and searching strategies should be a part of an introduction to online searching.

Critical Thinking Skills

Students need to evaluate the content of what they find online, just as they do with traditional information sources. Evaluating sources and comparing fact and opinion are essential activities when exploring online databases. Julie McGee of Ligature, Inc., describes a possible strategy:

> An elementary-school student has to prepare a report on George Washington. The student pulls out an encyclopedia, summarizes the article in a page, and turns it in. The student gets a grade. End of lesson—end of learning. Let's take a different approach. How could a teacher turn this exercise into a learning experience about both George Washington and information? One way would be to compare articles from several encyclopedias. What points does each stress? Does one contain facts that the others do not? The teacher could copy a page from a history book on the American Revolution, and the students could compare that information with the information in the encyclopedia. If the class has access to an online computer, an article from the electronic encyclopedia could be an interesting counterpart to the printed one. At the end of our revised lesson, the student would have acquired valuable tools for evaluating information, and the teacher could lead students into seeing the value of examining multiple sources of information and of knowing where to look for particular kinds of information. Importantly, students could begin to see that how information is organized can affect how it is perceived. (McGee, 1987, pp. 82–83)

DATABASES AND NATIONAL NETWORKS

Publishers of online databases are beginning to address the needs of the educational market and are offering attractive packages with special features geared to the academic community.

Einstein: The Information Access Tool, produced by The Learning Link, uses English commands and an easy-to-use menu structure. Einstein does not offer electronic mail or forums but provides over ninety databases which are selected by the publisher for their relevance to the curricula of elementary through high school. Einstein allows access to many different types of databases but translates the often complicated commands of specific services into one standard format. A teacher's manual with curriculum ideas and student activity sheets is available. One of the unique and advantageous features of Einstein is its cost structure. Single-session passwords are available, with special rates for large quantities. Because passwords can be purchased in advance and online charges do not accrue, districts with strict budget limits can budget funds for Einstein for the next fiscal year, knowing exactly what online searching costs will be.

Since 1976, *Dialog Information Services* has had a package for the school market called *DIALOG's Classroom Instruction Program* or CIP. Recently, DIALOG introduced *Classmate*, the Classroom Instruction Program for students from elementary to college. It includes materials to support an entire research skills curriculum. Although the full-service DIALOG offers hundreds of databases, Classmate limits the amount to approximately 35. These include databases in the areas of psychology, chemistry, government publications, and agriculture. *Books in Print*, the *Academic American Encyclopedia*, and *National Newspaper Index* which includes *The Wall Street Journal, The New York Times, The Christian Science Monitor*, and *The Washington Post* are available. Classmate can be accessed for $15 per connect hour, which includes the connection charges as well as the information, a self-instructional user's manual and a subscription to the *Classmate Newsletter*. Some of the educationally oriented databases in the Classmate package are:

Books

BOOKS IN PRINT: Currently published, forthcoming, and recently out-of-print books.

Business Information

ABI/INFORM: Business practices, corporate strategies, and trends.

TRADE & INDUSTRY INDEX: Indexes of major journals. Complete text of PR Newswire.

HARVARD BUSINESS REVIEW: Complete text of the Harvard Business Review since 1976.

Education

ERIC: Research Reports, articles, and projects significant to education.

Magazines

MAGAZINE INDEX: Index to articles in general-interest U.S. magazines.

Medicine

CONSUMER DRUG INFORMATION FULLTEXT: Complete text of *Consumer Drug Digest*

DRUG INFORMATION FULLTEXT: Complete text of the *American Hospital Formulary Service* and the *Handbook of Injectible Drugs*

INTERNATIONAL PHARMACEUTICAL ABSTRACTS: Research and current health-related drug literature.

MEDLINE: Biomedical literature and research.

News

NATIONAL NEWSPAPER INDEX: Covers *The Wall Street Journal, The New York Times, The Christian Science Monitor, The Los Angeles Times,* and *The Washington Post.*

NEWSEARCH: Daily index to over 2,000 news stories.

Reference

ACADEMIC AMERICAN ENCYCLOPEDIA: Complete text of the Encyclopedia.

MARQUIS WHO'S WHO: Detailed biographies on nearly 75,000 professionals.

IRS TAXINFO: Complete text of IRS publications.

MAGILL'S SURVEY OF CINEMA: Review articles on over 1,800 notable films.

An additional nineteen databases are available in the areas of agriculture, chemistry, computers, electronics, corporate news, engineering, government publications, law, mathematics, and psychology. Although the full-fledged DIALOG includes hundreds more databases in areas such as energy, the environment, and nutrition, the Classmate edition provides students with an enormous amount of information.

CompuServe is by far the most widely used system of its kind in the country. It is a virtual online playground with a wide variety of services and topics. It includes electronic mail, bulletin board areas for special interest groups, online shopping, and programs to download. The information available ranges from the weather in Africa to the price of fine wines from France. Information is updated frequently. A log-on sequence to CompuServe®, as with most of the other services in this section, is simple and quick (user responses are shown in boldface):

Host Name: **CIS**

User ID: **12345,6789**
Password:

CompuServe Information Service
06:39 EDT Friday 29-Jul-89

Last access: 15:03 28-Jul-89

Copyright © 1989
CompuServe Incorporated
All Rights Reserved

You have Electronic Mail waiting.

The caller is given a "What's New This Week" listing and then can choose to see the top menu or go directly to a particular section by typing a "GO" command, like "GO-AAE" for the Grolier Electronic Encyclopedia. Below is a copy of the TOP menu of CompuServe:

CompuServe TOP

1 Subscriber Assistance
2 Find a Topic
3 Communications/Bulletin Bds.
4 News/Weather/Sports
5 Travel
6 The Electronic MALL/Shopping
7 Money Matters/Markets
8 Entertainment/Games
9 Home/Health/Family
10 Reference/Education
11 Computers/Technology
12 Business/Other Interests

If the caller chooses "10," a menu is displayed that includes "Iquest" and "Academic American Encyclopedia." The AAE allows the user to type in a search term, and if the term is in the encyclopedia, see the full article on the screen.

Search term: **telecommunications**
Grolier
telecommunications

Telecommunications refers to long-distance communication (the Greek tele means "far off"). At present, such communication is carried out with the aid of electronic equipment such as the RADIO, TELEGRAPH, TELEPHONE, and TELEVISION. In earliest times, however, smoke signals, drums, light beacons, and various forms of SEMAPHORE were used for the same purpose (see SIGNALING). The information that is transmitted can be in the form of voice, symbols, pictures, or data, or a combination of these.

This article went on for several more pages. Online help is available in the form of command summaries:

B - Go BACKWARD a page
F - Go FORWARD a page
G 'n' - GO to page 'n'
M - Go to previous MENU
N - Display NEXT menu item
P - Display PREVIOUS menu item
R - RESEND page
S - SCROLL
S 'n' - SCROLL from item 'n'
SE 'n' - SEARCH for article 'n'

Another search service is available called *IQuest*. CompuServe describes the service as follows:

> IQuest gives you access to over 800 publications, databases and indices spanning the worlds of business, government, research, news—even popular entertainment and sports. Extremely easy to use, IQuest is a menu-based service which prompts you for your information needs and then goes to work for you. Accessing databases through online services such as DIALOG, BRS, NewsNet, and Vu/Text, among others, IQuest executes the search and displays the results to you.
>
> IQuest offers two simple methods for finding information of interest, IQuest-I and IQuest-II. IQuest-I guides you through a series of menus which define your topic of interest. Then IQuest determines which database is right for your search and prompts you for words to search for in that database. If you already know the database you want, IQuest-II allows you to specify the name of the database and bypass the menus.

IQuest allows you to access either bibliographic information or full text articles. The convenience of having the search performed for you is costly, however. Each search costs $9 and some carry a surcharge. Ten titles are displayed and if more have been found, an additional $9 is charged for seeing the next ten.

The Source®, another large, multipurpose electronic service, includes a vast array of functions including news, travel, online shopping, and special interest groups. *Today From the Source* provides users with news stories that are updated constantly as well as *Today in Congress* and *Opinion Forum*. Following is a sample of some of the numerous other services available, as indicated in The Source documentation:

> EMPLOYMENT SERVICE—an online listing of both candidates and jobs in 40 industry categories. Also available are tips on resume writing and interviewing, and editorials on headhunters and the job market.
>
> UNITED PRESS INTERNATIONAL—news, weather, and sports news via satellite from correspondents worldwide. Full text of the UPI wire searchable using key words, and stories are delayed only 90 seconds.
>
> ASSOCIATED PRESS—a keyword-searchable electronic edition of the AP newswire. Several hundred stories are specially selected by AP each day in categories ranging from "News Brief" and "World News," to "Business & Finance" and "Entertainment and Sports." Stories are archived for seven days.
>
> ACCU-WEATHER—domestic and international weather conditions and forecasts, provided by Accu-weather, a leading private weather forecasting firm. Regional highway forecasts are available during fall and winter months.

PROFESSIONAL EXCHANGES—designed around the information needs of people who use Personal Computers as part of their work. While SIGs provide a way for people with common interests to communicate, Exchanges offer indepth research capabilities, discussions and technical support for specific PC applications, as well as other valuable business information. Professional Exchanges include the Legal Exchange and the Educators Exchange.

These and many other areas provide a vast array of possibilities for classroom activities. The Source has a large section for educational purposes that includes the *Grolier's Academic American Encyclopedia.* A special feature of The Source is an online guided tour. There is unlimited opportunity to use the guided tour and it is free from online charges.

GEnie (The General Electric Network for Information Exchange) offers many of the same functions as the larger services and is less expensive than The Source or CompuServe. Available functions include electronic mail, round table discussion groups about various topics, and "live wire" real-time conferencing. Online multi-player games are a popular feature. There are no special membership offerings for school systems, but an education round table with a system operator is available.

Bibliographical Retrieval Service (*BRS*) offers the BRS/Instructor, a service "designed to help you give your students hands-on experience in database searching." The goal of BRS/Instructor is that of instruction, not research. There is a choice of a menu-driven or command format. The cost is $15 per hour, which pays for the access via Tymnet or Telenet, with no monthly minimum. Twelve passwords are given at the time of subscription. There are no royalty or citation charges when accessing information from over 100 databases. Free BRS/Search Service introductory training sessions provide instruction as well as written documentation. BRS/Instructor allows students to access a portion of the 145 databases available on the full-fledged version of BRS. These include:

Education

Education Testing Service Test Collection
ERIC
Exceptional Child Education Resources
National College Databank
Ontario Education Resources Information Database
Resources in Vocational Education
Vocational Education Curriculum Materials

Reference/Multidisciplinary

Grolier Academic American Encyclopedia Database
ACS Directory of Graduate Research (chemistry related)

Books in Print
BRS/CROS (Cross file searching)
BRS/FILE (BRS Database Directory)
BRS/NEWS (System Update File)
Dissertation Abstracts Online (Multi-disciplinary)
GPO Monthly Catalog (Government publications)
Knowledge Industry Publications Database (Publicly available databases)
Magazine ASAP IIItm (Full text magazine articles)
Magazine Index tm (General interest magazines)
Newsearch (Business, law, and general interest magazines, updated daily)
NTIS Bibliographic Database (Government reports, all areas)
OCLC EASI Reference (Online Computer Library Center, Inc. catalog)
Popular Magazine review Online (Index to popular magazines)
TERM (Social science thesauri)
Ulrich's International (Directory of Periodicals)
UMI Article Clearing House (document delivery information)

The above listings are only a sampling of the available databases in the areas of medicine, physical/applied science, life sciences, business, social sciences and humanities.

The *Dow Jones News/Retrieval* offers special rates for schools. For a fixed monthly fee, a school can get three passwords and unlimited usage. This is quite helpful when planning a budget and allows students and teachers to feel more comfortable using the system for longer periods of time. *The Educator's Guide to Dow Jones News/Retrieval* includes lesson plans for elementary through secondary grades. In addition to current events, lessons cover topics in the areas of mathematics, history, and biology. In Suffolk County, New York, the Third Supervisory District of the Board of Cooperative Services (BOCES III) has provided 72 school districts with the News/Retrieval service. Instructional consultant Frank Enright uses the service in his social studies classes when researching timely topics. He describes one experience:

> . . . . when the insider trading scandal on Wall Street broke at the end of 1986, it was not just some remote event to the children in our schools. It gave teachers a topical and current case study for exploring business ethics on a day-to-day basis. And talk about getting on-the-spot coverage: How much closer can you get than The Wall Street Journal? (Barrett, 1987)

In 1985, Apple Computer Company began offering *AppleLink* to its dealers so they could communicate with each other and with the corporate office. It is primarily used for information about policy, technical problems, and new product information. The ease of use is apparent in the main menu, which displays such icons as a picture of a phone representing the log-on function. It is only available for use with a Macintosh computer and is not available to any group of users except Apple dealers.

In the fall of 1988, *AppleLink®-Personal Edition* was created for general use. It works on a Macintosh®, an Apple IIe, and an Apple GS®. It is designed with educators in mind and offers such services as online instruction through Apple University, "chartered as the first electromagnetic, hyperspace, un-accredited university for users of Apple computers" (Apple advertisement). Courses include "BASIC Fun" and "Appleworks: Making Your Own Home Budget." There is also a "University of Tomorrow" with courses on SAT preparation and English composition. After enrolling in a class, students can take advantage of the school library, live interactions with teachers, and a student center. A calendar of events on AppleLink includes events in "The Auditorium" such as a visit with Steve Wozniak, one of the founders of Apple Computer. Unlike most nationwide services, AppleLink comes with its own software, making the need for telecommunications software unnecessary. The icon-driven main menu is artfully done and easy to understand. The representative pictures can be chosen by moving a mouse and clicking on the choice. AppleLink is available at $6 per hour and can be accessed through local numbers.

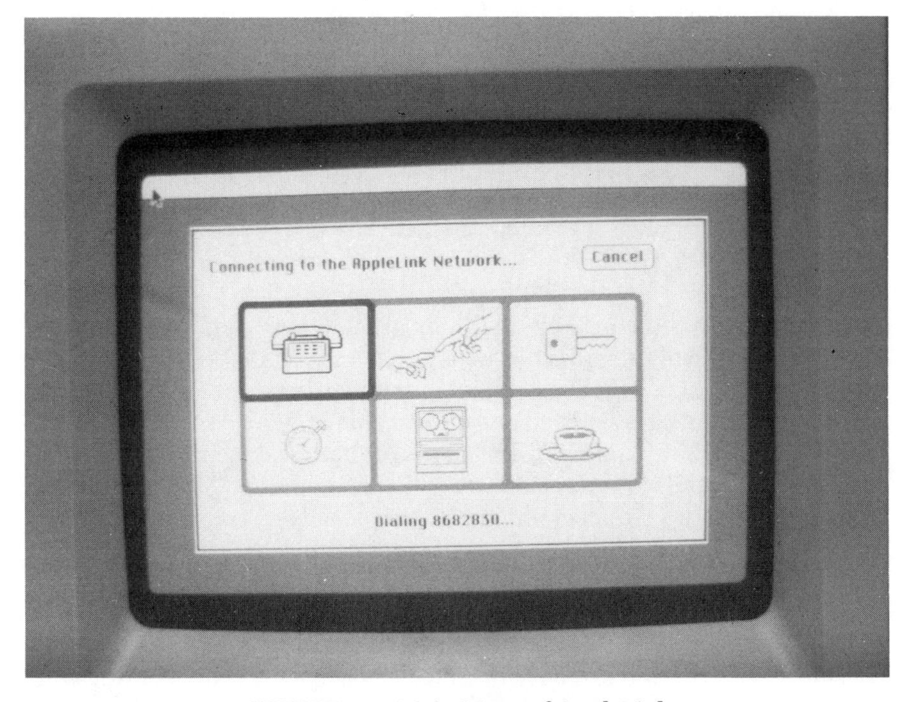

FIGURE 4.1. Main Menu of AppleLink

NewsNet, Inc. deals with primarily business-related topics. It includes 360 business newsletters which can be searched and scanned. A clipping service is available which allows a user to input a key word or topic. Names of appropriate articles are collected when the user is not online and a list is produced when the user logs on. The educational rate of $15 per hour is available for a package that includes nearly all of the functions. The commercial rate is $60 per hour with an additional $15 monthly fee.

VU/TEXT Information Service offers full text of articles from magazines, wire services, and forty local and regional newspapers. It also includes the Grolier's Academic Encyclopedia. Although the usual rate is over $100 per hour, a generous educational discount brings the rate to an affordable $6 per hour plus an additional $10 per hour for Telenet or Tymnet access.

Prodigy Interactive Personal Service, a recent project resulting from a partnership of Sears, Roebuck & Company and International Business Machines Corporation, is a local service that has gained popularity in the Atlanta and San Francisco areas and is moving to other cities around the country. It offers a variety of services such as online shopping, electronic mail, weather, news, sports, and stock information. It has an online version of *Weekley Reader* as well as educational games. It is similar in many ways to Boston CitiNet, mentioned in Chapter 3, in that it is geared to home consumers and is supported mostly by income from advertising. The start-up kit costs $49.95 and there is a monthly charge of $9.95. Although its use is limited to IBM and Macintosh computers, its low cost and the services offered make it an attractive package for schools.

CARRIER SERVICES

Carrier services allow users to call a local telephone number in order to access a host computer that is in another city, state, or even another country. The cost is usually absorbed in the hourly fee for the out-of-town service but sometimes is listed as an extra charge.

PC Pursuit, offered by Telenet, is a unique service that allows unlimited use of telecommunications for a fixed fee of $25 per month. Twenty-eight cities are accessible by the service. By calling a local number and typing in a special password and access code, users can then connect to a bulletin board system in another city without any other charges. There is no additional charge when used during evening and weekend hours, but there is a surcharge if it is used during the business day. It still offers a significant savings and can be helpful when a class is communicating with another class via a local bulletin board system in another state.

QUESTIONS WE MUST ASK OURSELVES

With the increasing amount of information available, it is easy to lose sight of its worth and purpose. Alex Molnar states:

> An increasingly difficult problem for educators: how to make sense of and respond to the increasing volume of information available to them? This problem has both policy and curriculum and instruction aspects. (Molnar, 1986, p. 64)

How can children learn to evaluate the information they obtain? Children often take the information from a computer as fact, without evaluating it as carefully as they would the written word. Activities that involve distinguishing fact from opinion and evaluating and comparing sources are essential when using online information.

Just what kind of sociological implications does telecommunications have? Sherry Turkle, in *The Second Self: Computers and the Human Spirit*, discusses the political and social ramifications of personal computers and the developing computer culture.

> Personal computers were small, individually owned, and when linked through networks over phone lines they could be used to bring people together. Everything was in place for the development of a politically charged culture around them. The computer clubs that sprang up all over the country were imbued with excitement not only about the computers themselves, but about new kinds of social relationships people believed would follow in their wake. Of course there was talk about new hardware, new ideas for programming and circuit design. But there was also talk about the rebirth of ideas from the sixties, in which, instead of food cooperatives, there would be "knowledge cooperatives"; instead of encounter groups, computer networks; and instead of relying on friends and neighbors to know what was happening, there would be "community memories" and electronic bulletin boards. Computers, long a symbol of depersonalization, were recast as "tools for conviviality" and "dream machines." Computers, long a symbol of "big"—big corporations, big institutions, big money—began to acquire an image as instruments for decentralization, community, and personal autonomy. (Turkle, 1984, p. 172)

The quantity and value of online information has been demonstrated. The opportunity for teachers to enhance their classroom curriculum is here. What remains is the responsibility to help students search for information in an efficient way, to evaluate what they find, and to use it to extend their knowledge.

REFERENCES

Barrett, M. (1987, July/August). Reading, 'riting and 'rithmetic: News/Retrieval adds fourth "R" for 10,000 students in New York. *Lines to Learning, Dowline.*

Cook, P. (1987). Multimedia technology: An encyclopedia publisher's perspective. In Sueann Ambron & Kristina Hooper (Eds.), *Multimedia in education, Apple Education Advisory Council: Learning tomorrow* (pp. 237–262). Apple Computer Co.

McGee, G. (1987). Curriculum for the information age: An interim proposal. In Mary Alice White (Ed.), *What curriculum for the information age?* Hillsdale, NJ: Lawrence Erlbaum Associates.

Mendrinos, R. (1988). Speech presented at Lesley College.

Molnar, A. (1986, March). Common sense about information and information technology. *Educational Leadership,* pp. 64–65.

Schwartz, J. L. (1987). Closing the gap between education and schools. In Mary Alice White (Ed.), *What curriculum for the information age?* Hillsdale, NJ: Lawrence Erlbaum Associates.

Turkle, S. (1984). *The second self: Computers and the human spirit.* New York: Simon and Schuster.

PART II

SCHOOL APPLICATIONS

5

Language Arts, Bilingual Education, and Foreign Language Teaching

Now that the wealth of potential resources and how to access them has been presented, what does all of this mean to the teacher and student? Part II of this book describes a host of current, ongoing telecommunications activities found in classrooms across the country. These activities allow students a deeper and more immediate understanding of their world as well as enhance and enrich learning in the various subject areas.

The discipline area called language arts has multiple facets in today's schools. Language arts teachers are responsible for teaching reading, writing, and literature. These broad categories encompass a wide range of skills such as reading and writing in all other disciplines, factual knowledge in subjects such as grammar and spelling, and appreciation of literature, drama, and poetry. This multitude of goals can be overwhelming unless the teacher is very creative and open to a wide range of devices and strategies. Telecommunications is a language arts tool that gives teachers another strategy for increasing their students' communications skills. This chapter looks at three areas of language instruction—English language arts, bilingual education, and foreign languages—and how telecommunications can augment the teaching of each.

TEACHING OF ENGLISH LANGUAGE ARTS

Language is a process for communicating ideas. To separate the different elements of language for the sake of teaching does not make sense. The National Council for the Teaching of English states, "Teachers and researchers now emphasize more than ever the interrelatedness of all language abilities: speaking, listening, reading, and writing" (NCTE, 1986, p. 3). Using telecommunications in language arts facilitates the integration of these language areas. Let's look at a writing lesson as an example.

> A junior high school teacher decided to use the school district's electronic network to motivate creative writing. She wrote the beginnings of a story and distributed it to her class. Together the students brainstormed next possible scenarios. The teacher then sent the story beginning and the several suggested scenarios to another junior high school English class on the network. Their task was to write the next component of the story, do their own brainstorming session, and put the extended material on the network for a third class. After a few iterations, the original teacher made hard copies of the current story version, passed it out to each student who had been involved, and asked each to complete the story. A best story was selected by a panel composed of students from all the classes and the winning story was published in the school district's literary magazine. (Oklahoma State University Teleconference, 1988)

The unique element of this exercise is that every student in the classes involved had an input, in some way, to the chosen story, and students used all the language skill areas during the activity. By using the network, the project was completed in a timely manner before interest was lost.

Getting Started

A very common way to introduce telecommunications into the classroom is through pen pal letters. It is relatively "low tech," since all the class needs is a computer, a modem, the appropriate software, access to a telephone line, and access to another class that is similarly outfitted. Compatible software and a single (rather than switchboard) telephone line help keep the technical worries to a minimum (see Chapter 2). The two teachers involved should work the bugs out of the system by communicating with each other several times before trying it in class.

Several different ways to motivate the exchange of letters have worked successfully for a wide variety of ages.

- The "Santa Claus" project, developed by Al Rogers, creator of FrEdWare, matches an elementary class with a middle school class, the latter playing Santa Claus (Dodge & Dodge, 1987, p. 18).

- The "Great Pumpkin Letter Writing Campaign" was devised by a seventh-grade teacher with a class of writing-resistant students. She paired her class with a class of fourth-graders in another school. These letters took the form of a big brother/sister relationship which worked to encourage the writing of the seventh-graders (Schrum, *et al.*, 1988).
- In an isolated Maine community, Dennis Dorey wanted to increase the amount of communications his students had with people of the same age. Using MeLink (see Resource Section), he had his students write to another class of twelve-year-olds (Dorey, n.d.).
- Two Maryland teachers of gifted students started a telecommunications project initially using pen pal letters.

Without a focus, however, pen pal letters are hard to sustain. Further, how much learning about writing actually occurs becomes questionable after a while. Students do need to know something about the people with whom they are communicating before they become involved in specific projects.

Peter Hutcher and Joan Winsor created guidelines for the first pen pal letter exchange that seems to give enough structure to the activity to ensure success. The student is presented with the following outline:

> This form will help you introduce yourself to a new pen pal.
> Answer each question with at least one sentence.
> Remember to use complete sentences.
> Begin your letter with "Dear pen pal," and skip a line.
> What is your name?
> What is your birthday?
> Where were you born?
> How old are you?
> What grade are you in?
> How many people are in your family?
> Who are they?
> What is your favorite subject in school?
> What is your least favorite subject?
> What is your favorite and least favorite color?
> What is your favorite and least favorite TV show?
> Do you have any pets? What are they?
> What do you like to do? What are your hobbies?
> End your letter.
> Say goodbye and write your name.

Joan Winsor comments on using this template:

> This summer Peter and I started out using a FrEdWriter prompted file for the kids to use in constructing their first letter. That way they all introduced themselves by providing the same background info. When my kids read the ones that Peter sent, they were very interested. One kid found someone who he was dying to write to because that person (a girl, even!) liked everything he did—but someone else had already taken her. He was sooooo disappointed! I think that following some sort of pre-planned format is definitely helpful, and makes for more interesting first letters. After that, they have something to talk about. (Hutcher, 1987)

The Maine and Maryland exchanges both developed into more integrated language arts projects. The Maine teacher next gave his students the assignment of writing a biography of their pen pal. First the class brainstormed about what sort of information is needed to write a biography. The students then gathered the recommended information through their pen pal letters, wrote the biographies, and uploaded them to MeLink. The challenge then became to identify each student. The culminating activity was a face-to-face meeting of the two classes at the end of the school year.

Following the initial contact, the Maryland teachers matched their students based on research interests in such things as history, geography, careers, and local industry. The sharing of local information in the form of research reports became part of the social studies curriculum. "With peers as an audience, students felt freer to commit their thoughts to paper and more responsible for the historic information they recounted" (Baer, 1988, p. 22). The language arts part of the telecommunications activity was similar to the Maine project. Here the students told their pen pals about themselves using pseudonyms. The culmination was a two-day meeting where students sought out their pen pals.

Working with severely emotionally disturbed high school students in the Montgomery County Intermediate Unit, Norristown, Pennsylvania, Fred Wheeler set up a bulletin board to motivate his students "to write more frequently as a means of increasing their language arts." Wheeler comments:

> The motivation to use telecommunications is not just to work on language arts skills but to add to the affective areas of the student's learning. The real success of this project cannot be measured by just counting the number of times the BBS is used or the number of spelling and grammar errors in a student's writing. Although language arts skills have increased, the greatest increases have been in the affective area. As affective goals are the most important goals in our students' educational plans, this has been extremely gratifying.
>
> For some of our students, communicating through the bulletin board has been the only area in which they reacted and interacted in socially acceptable ways. The benefits in terms of increased confidence, improved self-image,

> and increased ability to deal with frustration have been immeasurable. To see a student, who once felt too "dumb" to let anyone see anything he wrote, enjoying the messages he sends and receives is immensely rewarding. (Wheeler, n.d.)

Dawna Traversi, a special education teacher in a Cambridge, Massachusetts, school, introduced telecommunications to her resource room students by conducting a sports survey using CitiNet, a regional network. The students put the gathered information on AppleWorks and then analyzed it and wrote reports. Dawna, however, found a conflict between online writing and maintaining correct spelling and grammar. In addition, she was not sure which curriculum objectives she was meeting (Traversi, n.d.).

Meaningful Communications

The National Council for the Teaching of English (NCTE) observes that:

> Many English language arts teachers now agree that by experiencing and using language in a variety of contexts, students achieve personal growth: they respond to their experiences and learn about their world, their feelings, their attitudes, and themselves. (NCTE, 1986, p. 3)

An area that elementary students are anxious to know better is the world they are about to enter in their "move up" to middle or junior high school. Several teachers have designed telecommunications projects to relieve the anxiety of their elementary students. This next project was developed by the language arts teachers and the computer curriculum coordinator of five rural western Massachusetts towns.

> Classes were paired, then students within each paired class were given pen pals. The students' first task was to create a design, using the Logo computer language, that represented them for their pen pal. The second task was to write to their pen pals using the Logo designs as a basis for conversation. Before the electronic letters were written, the language arts teacher discussed with each class "the art of non-contact communications . . . keeping in mind all the tips important in telecommunications—short and unified messages, when answering refer to the topic, end with a question, convey your feelings." (Monroe, 1986, p. 3)
>
> Students then practiced this kind of communications in the computer lab with each other, using splinter cables and two monitors each, one from their computer and one from a classmate's computer. This allowed the students to compose messages and then see the answer from their partner's computer. "These activities help students understand that messages need to be brief, clear and accurate, espe-

cially because they are made without the aid of body language or the ease and speed and inflection of speaking." (p. 4)

During the next language arts class, the students wrote their autobiographies. In the computer class they telecommunicated these to their pen pals using pen-pal nicknames.

The second part of this project was telecommunicating with the seventh graders. Students asked questions about drugs in the junior high school and being bullied, and received information from their older schoolmates about these concerns. Once again the culminating activity was move-up day, when the sixth graders from the five separate schools met both each other and the seventh graders at the junior high school.

The success of this project is demonstrated by the number of follow-on projects. The network is now being used by the students to write editorials about issues common to all five towns such as garbage dump problems, budgetary debates, and town growth issues. They are using the network to share experiences from the Outdoor Week program in which each class participates. The project being planned for next year is to "talk" with students in a New Jersey town about the problems of small resort towns.

Reil used telecommunications as a newswire service to "explore more fully the influence of 'audience' on students' writing and revision" (Reil, 1985, p. 323). For the purpose of this study, The Computer Chronicles Newswire was created and linked together rural and city schools in Alaska and California. A key element of the project was the Editorial Board which had to choose the articles from the newswire for their school newspaper. In the process, the students had to decide what determines "good" writing, what makes an article newsworthy, and what is and is not in good taste. Students became so enraptured by this project that they eagerly gave up their recess time to participate in it. Reil concludes:

> The students that we worked with experienced a sense of power and control over the medium. The computer was a tool that they used to help them share life experiences of children who were living in a world very different from their own. Writing and reading, editing and revision became means to serve this goal. The students helped one another and received help from the computer program and the computer coaches. But . . . their newspaper had become their own. (Reil, 1985, p. 335)

Not all initial attempts at getting started work out as well as the above experiences suggest. Several teachers interviewed mentioned it took an unreasonable amount of time, sometimes approaching a full year, to get telecommunications working. The major reasons given for the problems were lack of funds, training, and technical support: "Most schools do not have in-house technical people who can cope with the complexities of telecommunications and keep systems going" (Eliot, n.d.). In spite of the many complaints, however, the majority of teachers stick with the project until they are "up and running." These teachers become very dedicated to the potential of the technology and begin doing more and more ambitious projects.

Other Writing Projects and Some Evaluations

According to the NCTE:

> English teachers . . . must be able to create situations in which students discover the importance of language and gain skill in its use. Such situations will require that students have a variety of opportunities to use language. Actively use language rather than passively listen to the teacher talk about it, is the essence of much student learning in the English language arts classroom. (NCTE, 1986, p. 13)

Cohen and Riel conducted a study to see whether students would write better compositions when they were writing to their teacher or when they were writing to other students. Two seventh-grade classes were asked to write on the same topic. One composition was to their teacher for their semester grade and the other composition was to a student in another country as part of a telecommunications project. The teachers were asked to predict which composition would receive the higher grade. Both teachers chose the composition that was written to the teachers for the semester grade. However, in fact the grades of the composition to the other students were higher (Cohen & Riel, in press).

For an elementary class in Pennsylvania, talking with fellow students in Australia highlighted cultural and geographic differences and similarities. Immediate differences were noticed in each student's use of the English language and the respective country's geography. Comforting similarities were found in such things as their preferences for sports and TV shows. The first year of this project had three phases: getting acquainted, acting as resources for each other, and collaborating on a project—jointly planning a Halloween party.

> The Australian children wrote and asked how Hallowe'en was celebrated in the US; what the children would do, what they would eat, what games they play. Answers came back with suggestions on how Hallowe'en may be celebrated.
>
> A day of celebration was planned based on the replies. The children all made costumes. Each prepared some special food for the day. We bobbed for apples, played tricks on the other classes, had a special lunch and went online to send messages to the children in Pennsylvania. After the day the children wrote about what we had done and how we had celebrated Hallowe'en. In return we received reports on how Hallowe'en had been celebrated in the US. (Butler, n.d.)

> The emphasis was on clear communications with other students, so both writing and reading had a real purpose. And the almost instantaneous correspondence provided by telecommunications was highly motivating for the students, writing to each other halfway around the world. . . . Students developed more effective written communications skills, exchanged cultural information and became familiar with an up-to-date technology, in one integrated activity. (Butler & Jobe, 1987, pp. 25–26)

The following comment resulted from a similar activity linking an American class to an Australian class:

> It is not the technology that will cause a project to be a success or failure, but rather the willingness of participants to invest themselves in a process that takes infinite patience to coordinate [schedules, vacation times, time zones]. When teachers are able to cooperate with colleagues they never met who live thousands of miles away, they are teaching cooperation in the best way possible: by modeling! Telecommunications is a resource that allows us to move students into a world where cooperation is becoming increasingly important. (Lake, 1988, p. 19)

Fred D'Ignazio puts meaning into students' work by creating "multi-media classrooms." These classrooms have access to three media centers housed on rolling carts. One is a video cart containing a VCR, video camera, and small TV. The second is an audio cart with a boom box, electronic keyboard, and boxes of audio tapes. The third is a communications cart containing a computer, modem, printer, and appropriate software. The students create and teach their own lessons. As part of the communications aspect of the multi-media classroom, Fred has connected, through the Imagination Network, thirteen Alabama classes with thirteen British Columbia classes.

> We call the Network 'the classroom without walls.' When computers hook up, it's as if the classroom walls come tumbling down, and thousands of miles shrink to nothing. Canadian students send stories of what it's like to live on Vancouver Island. Kids at Cahaba Heights send tales about life in the heart of Dixie, about Bama football, storytelling on the Selma riverfront, and Mardi Gras celebrations in Mobile. The Imagination network is a two-way multimedia highway, where distance is only a state of mind. (D'Ignazio, 1987b, p. 11)

> Students and teachers have come to know each other on a one-to-one basis. They have set up pen pal networks and "special interest groups" on pets, model building, cars, music, astronomy, math, and many other subjects. They also collaborate on many joint projects, including newsletters, research projects, electronic stories, science projects, video production, and electronic music. Reading and writing have become vastly more meaningful to the students because they have someone to communicate with—they have a reason to communicate. It is so exciting to be in close touch with a fellow

> student who lives far away in a distant country. It is fun to learn about their world and tell them about your world. (D'Ignazio, 1987a, p. 23)

Nina Myers, the English and Computer teacher at the Miller Junior High School, San Jose, California, has combined her two subject areas in some unique ways:

> A colleague at a sister school bought a modem this academic year and our journalism students have exchanged school news for our respective papers.
>
> When each of us has had distinguished visitors, the (unvisited) school called the class on the modem and chatted with the visitors. Two cases in point which caused considerable interest were when Hyde School had Russian visitors and we chatted with them and tried to use some Russian words in our greeting. When we at Miller had Japanese visitors Hyde School chatted with our group. (Myers, n.d.)

Nina also commented on how much her students enjoy the "chat" function of their bulletin board. Students really are thrilled with the direct contact they have with other students whom they might never meet. To encourage students to use the electronic mail system, Nina selects two top students each year from her programming class and has them become the system managers of the bulletin board. They read the manuals, set up the mailboxes, and teach the other students how to use the system.

When taught from the process approach, writing usually involves at least three steps: prewriting, when ideas are generated, details collected, and materials organized; the first draft, when the ideas are put together into a coherent piece; and the rewriting, when the manuscript is polished. Bruce Fleury, a sixth-grade teacher in San Diego, integrated telecommunications into his writing program to specifically support writing as a process. He has elicited the help of retired writers and utilizes their expertise through a network. His students' writings are sent, overnight, to the professional writers. They comment on the compositions, online, and send their comments, via the network, back to the students. The cycle continues until the teacher and students are satisfied (Davis, 1987, p. 67).

As a result of a grant from Tandy Company, Carmel High School in Carmel, California, integrated telecommunications into its writing program in a somewhat different way. The Tandy grant consisted of a number of laptop computers as well as a bulletin board system installed on a larger Tandy computer. Students wrote their papers using the laptop computers. When they were ready, the students uploaded their work for their teachers to the bulletin board on the larger computer. At their leisure, the teachers retrieved the papers, commented on them, and sent them back to their students. This became an iterative process, taking place without the usual crunch of deadlines producing large amounts of papers for the teacher to correct. Both students and teachers felt the electronic dialogue encouraged better teaching and learning.

Feeding the Billions, an interdisciplinary project designed by language arts teachers, social studies teachers, and the library media person at the Thurston Junior High School in Westwood, Massachusetts, attempts to help students understand the problem of hunger in third world countries. The language arts teacher's role is to work with the students in writing an essay around a fictional character from a third world country. The information in the essay is based on the facts gleaned from their social studies work using online databases for up-to-date data (see Chapter 6). In addition, the students communicate, through AppleLink®, with schools across the country to survey people who have recently emigrated from third world countries to the United States. Roxanne Mendrinos, the library media specialist, concludes:

> The students are highly motivated to learn. Their writing has improved, including their use of adjectives and descriptive phrases. They are far removed from plagiarism. Students research, format their information in outline form and in spreadsheets, analyze the facts, and make projections before writing their first draft. The students are creating historical fiction by using their research and superimposing themselves into family life based on these facts. (Mendrinos, n.d.)

Students themselves can see the improvement in their writing and their willingness to write, as illustrated by the following statements by some of the Jefferson Junior High School students from Oceanside, California:

> Before this project, I did not like to write. Now, most of the time my teacher reads my papers out loud in the class.—G. Ruiz

> I find that knowing the things I write are going to be read by people everywhere, helps me be more critical of my writing.—T. Lee

> Ever since I started writing on the computer, and sending my stories other places, my compositions have been getting better.—B. McGeehee

> Right now we are doing a telecomm project and I love it. It gives you a chance to brainstorm and read the works of kids your age from all over the world. It is just great!—J. Carter (Andres-Syer, 1988, p. 8)

In summarizing the writing projects done using the Computer Mail System (CMS), active on the west coast, Tony Anderson has observed:

> . . . never-ending stories done via bulletin board sections, riddle collections, magazine publications using materials gathered via the network, Santa letter writing, pen pals, use of Kidwire by journalism students, and dissemination of FrEdWriter-prompted writing lessons. Other kinds of projects include: a collection of student-written book reports gathered into an AppleWorks database; the operation of a Kid Travel Agency in Nebraska, using local attractions across the country; junior high students writing to sixth graders

> about their concerns and fears of moving on; technical help via online assistants . . . and brain teasers/trivia contests conducted online. (Anderson, 1988, pp. 10–11)

Sounds like the contents of CompuServe or The Source!

If a media specialist is not available, library research strategies might be the responsibility of the language arts teacher. According to the NCTE:

> Teachers must be constantly attuned to both immediate and long-term effects of social issues and world events. This sensitivity enables teachers to link vital issues and events with the goals of English language arts instruction. Such connections between school and the outside world sustain students' motivation to learn. (NCTE, 1986, p. 16)

An example of how exciting database searching can be is relayed by Lee Sapienza, a Wayland, Massachusetts high school teacher:

> Last fall I witnessed a scene that can only be described as an educator's dream come true. A young girl, seated at a terminal in the computer room of our Media Center, was becoming visibly excited—a broad grin of delight slowly spreading across her face. Finally, clapping her hands together for emphasis, she cheered, "I LOVE IT!" The "it" to which she referred was not the image of a Bruce Springsteen or a Michael Jackson but the steady scrolling across the computer screen of *The Electronic Washington Post,* full text, as it downloaded from Dialog Information Services in California. The student was researching the use and effects of Agent Orange for her Honors Chemistry class, and had really hit pay dirt in the online databases that morning. (Sapienza, 1988, p. 5)

Larger Projects

Quill, begun in 1981, developed software to support "writing as a process" and is one of the earliest writing projects to make use of networking. The three aspects of the *Quill* software are: the Planner, for organizing ideas on a specific topic; the Library, for storing draft copies and finished documents; and the Mailbag, an electronic mail and bulletin board system. After teachers were introduced to Mailbag, it

> . . . became the basis for frequent writing activities. Students enjoyed writing and sending messages, and posting them in both the computer bulletin board and bulletin boards in their classrooms. As the year progressed, teachers reported fewer problems with typing skills. . . . But as students' typing and writing ability improved, the demand for computer use was even greater perpetuating the frustration caused by the availability of only one computer. (Bruce & Rubin, 1984, p. 23)

Mailbag is used to motivate students to try different kinds of writing. The themes of pen pal letters are constantly changed so that students try humorous, philosophical, informational (trip information, game information), direction giving, and persuasive writing (persuade your pen pal that historic novels are fun to read) over the course of the year. The researchers from the *Quill* project find:

> The Mailbag enhances writing instruction by:
> - Encouraging written communication to varying, but specific, audiences (for example, friends and classmates).
> - Allowing different kinds of writing to occur (for example, informing, persuading, instructing, entertaining).
> - Motivating students to write more by personalizing the experience. (Bruce & Rubin, 1984, pp. 10–11)

In addition, other benefits of *Quill* are noted. One of the student's fathers made an appointment with the school administrator "to thank her for giving his shy daughter the opportunity to open up. The child never felt comfortable talking in class, but interestingly, the Mailbag had begun chipping away at her shyness. She proudly announced to her father that she and her teacher wrote messages to each other all the time" (Bruce & Rubin, 1984; Appendix A).

Quill as just reported consisted of electronic mail essentially within single classrooms. The second part of the project, carried out in Alaska, was more ambitious. Several Alaskan classrooms were networked for enhancing writing experiences. Because the project was done in the early years of networking, one teacher had to physically transfer files from one network to another for the messages to reach their destination. This person became the "gatekeeper" of information for the *Quill* network, creating and sustaining conversations throughout the state. As a result of this technical function, the network, originally set up for student-to-student and teacher-to-student communications, became equally important for teacher-to-teacher communications (Rubin, in press).

BreadNet is an outgrowth of the Bread Loaf Program in Writing to improve teachers' skills in the teaching of writing, held each summer at Middlebury College in Vermont. As a way of extending the Bread Loaf experience beyond the summer, *BreadNet* has been established to allow alumni to experiment with telecommunications as an aid to their teaching of writing. The teachers have found that using electronic mail rather than the Postal Service eliminates one step that can discourage teachers from this kind of sharing:

> Think of how much easier—and thus more likely to happen—it is to sit down at a keyboard and immediately send a message than it is to write a letter (or bundle up a dozen of students' letters), find a stamp, make time to go to the post office, and wait days for a reply. (Walker, 1987, p. 14)

BreadNet has many elements in common with the other networked writing projects. As with the Alaskan *Quill* project, *BreadNet* is as essential to teacher communications as to students. Teachers continue their discussions about teaching writing, begun with their summer Bread Loaf experiences, online. One teacher described how his students did revisions using the word processor, and raised several questions about directions he should go to improve their skills:

> How else could an overworked teacher in a remote area discuss writing theory and practice with interested colleagues in the middle of the school year? (p. 14)

A second element, found both in *Quill* and *BreadNet,* are people who read all the conversations on the network, monitor them so as to link people with common concerns, offer encouragement to teachers experiencing difficulties, and generally contribute to the online discussions. In the case of *BreadNet,* these "gatekeepers" are faculty from the Bread Loaf program.

As with several of the previously reported projects, students on *BreadNet* first need to get to know each other before plunging into more academic projects. One interesting combination of students came about when teachers from a small western ranching town, an Indian reservation, and an eastern independent school decided to telecommunicate. The project started by students interviewing a classmate and then introducing that classmate to the network. Following this, students wrote autobiographies online. Next came more in-depth descriptions of life in the diverse communities. Styles of writing, including humor and sarcasm, begin to emerge. The following is an exchange between the ranch town and the independent school.

> Some have said that, being in a town of 200, and being 40 miles away from any town of reputable size, we are secluded, and at times we have been considered hicks. This is not so. Few of the kids wear brogans and straw hats to school, and even fewer ride mules. Sneakers, buttondowns, and Wranglers are quite prevalent among the students, most of whom are quite conservative. We do, however, have a small new-wave factor, which is definitely in the minority. Most of the kids do tend to leave their horses home on Saturday nights, preferring hot engines to cold saddles. I hope I have now clarified our position in society for y'all in the big city. (p. 15)

A question about Gucci prompted the independent school students to write:

> I told my brother Biff, and my other brother Skip, and Muffy my sister, and Mums, about how none of you had heard of Gucci. They were so sympathetic that they have organized a relief fund—Aid for Gucci. To raise money, everyone here at school has donated all their old pink shirts, light green shirts, khaki slacks, Izod jerseys, scarfs, Blouchers (the preppie shoe), and old lacrosse sticks. (p. 16)

The tone of the contribution from the Indian Reservation school is somewhat different:

> Problems facing the Lakota people are the abuse of alcohol and drugs. Even with the low economy a high percentage of people use them, with alcohol being recorded as the number one cause for deaths. Drugs are not considered a killer but are highly used.
>
> The life outlook for the Lakota people is going in a downward direction, because of the high rate of deaths and abuse. To control it would take great human power or magic. . . . (p. 16)

This narrative led to an extended unit to better understand Indian problems. Federal aid to Indians and history from the Indian perspective were studied. Again, exchanges that are initially designed for getting to know people almost always end up with extended study that augments the curriculum.

> As a final assignment in the course, the students wrote about what they had learned from participating in the three-way exchange. One student was moved to write: Something that is strong in my mind is a kind of respect and understanding that we have built with other students through writing. . . . At first it was shocking to hear about what was going on around them. But after we shared our differences, I feel it made us understand each other better. . . . The understanding came because we were sympathetic and curious about their environment and they were willing to make us understand. . . . I guess writing took over as our source to communicate and get to know these distant people. (p. 16)

The *BreadNet* teachers are also having success using the network with special education students. Learning disabled elementary students are paired with college students who had also been diagnosed as learning disabled. A fourth-grade student began with the message:

> have you ben lurny alot at school I have in math .math is my frate subeget. I downt know way. Wought is yours. I wortht it By my saif . I know I am not the best in spelling .I not a fast Lurrn. Are you I bet you are.

At this point the student becomes frustrated and the teacher types while the student dictates.

> It's no fun not knowing how to spell words. When I came to this school I did not know any of my times tables or my minuses or pluses or my long division. . . . Now I'm catching up with the fourth graders. I'm always like one page behind.

The response from the college student is:

> I'm very glad that you typed some of your letter. I know it's hard. I learned to type in the 10th grade. It took me a long time to learn how. I didn't like it at first, but it became easier the more I practiced. That's the way most things are.

> I know that spelling must be hard for you. It's the same for me. Don't let it get you upset. You're young, and you'll have many many years to learn everything that you need to know. I used to get mad when I didn't understand something. I used to think that I wasn't as smart as the other kids in my class. . . . Now I know that I was wrong.

The correspondence is most meaningful, as documented by this response from the fourth grader.

> Hi how are you . . . I like your letter. I'm glad for what you have to say. It helped me a lot. I look at it practically every day. Did you write it by yourself. I'm sure you did. I wrote a little bit of this by myself and then Ms. Tulonen helped me finish it off while I told her what to say.
>
> I have enjoyed these months writing to you a lot. It has been very interesting for me because sometimes I get to skip my class to do it. Talk to you later. Keep writing. Love, Jenny. (p. 18)

BreadNet has not formally evaluated its projects. However, one of the teachers "tentatively concluded that students who had frequent access to computers for word processing did more kinds of writing than those who wrote by hand or had only occasional use of computers. Add to that a real audience provided by the telecommunications link and you have a powerful tool for getting good, engaged writing out of students" (p. 18).

Again, the going has not been all smooth for these projects. The Bread Loaf program focuses on improved teaching of writing. Networking is only viewed as a possible tool to that end:

> Weaving the *BreadNet* that is making this collaboration possible hasn't been easy. Getting used to unfamiliar equipment, encountering glitches in the system, and figuring out how to use the network in a meaningful way in the context of a classroom all take patience. (p. 20)

Another project whose focus is primarily writing, is Computer Pals, started in 1983 by Malcolm Beazley, head of the English Department at Turramurra High School, Syndey, Australia. The project now links together classes all over the world, but initially began by linking schools in Australia with each other and with schools in Alaska. The aims of the project are to:

1. Provide a real context in which students can improve their written communication skills.
2. Provide an opportunity for cultural exchange through writing and related language arts activities, with a view to developing greater understanding among people irrespective of race or creed.
3. Encourage the less linguistically motivated student.
4. Provide an opportunity for students to develop keyboard skills in a purposeful context.

5. Familiarise students with the use of international tele-communications. (*Computer Pals Newsletter*, 1988, p. 3)

An interesting aspect of this project is the requirement that every participating classroom follow a set pattern of interactions. The project leaders feel this ensures the success of the experience. These stages of interactions are:

> *Stage 1.* Getting to Know Each Other—This stage is very similar to the pen pal prompts set up on *MIX* by Peter Hutcher.
>
> *Stage 2.* Reporting on Matters of Everyday Interest—Students write reports on subjects that are of interest to themselves. For example, an Alaskan student did a report on the Aurora Borealis, while an Australian student did one on the geography and culture of Australia.
>
> *Stage 3.* Expressing Oneself Through Poetry—"Exchanging poetry with people from a vastly different culture encourages a lively new interest." (*Hands on Around our World*, n.d., p. 4)
>
> *Stage 4.* Presenting the News: A Lesson in Electronic Journalism—For this stage, a monthly newsletter is produced, with each child being responsible for a different topic.
>
> *Stage 5.* Discussing Social Issues—A current issue is selected and students carry on an online discussion.
>
> *Stage 6.* Exploring our Cultural Heritage Through Theatre—"The final stage of this program encourages students to write scripts on the myths and legends peculiar to their culture. . . . The scripts can be used to develop dramatic activities in the classroom through puppets and many other forms of theatre." (p. 4)

A very interesting concern is posed by A. P. Scott, one of the teachers from Great Britain who is part of the Computer Pals network. Scott questions the teacher's role in telecommunications:

> A message from a UK primary pupil to an Australian primary pupil might well go from child to child, but is in fact much more likely to be a group effort supervised and dispatched by a teacher to another group, received and also supervised by a teacher. . . . Messaging is going to be mediated by teachers, trained in their "own" culture. In a "conversation" between Italian and Alaskan pupils, this is likely to make a considerable difference. One is entitled to suspect that a pupil sitting in front of a terminal reading and dispatching material to another pupil sitting in front of another terminal is a rare event. (Scott, 1988, pp. 11–12)

This also has implications for the notion of a gatekeeper to monitor a network's activities, which seems to work well for *Quill* and *BreadNet*.

The InterCultural Learning Network, based at the School of Education, University of California, San Diego has many goals in addition to enhancing student writing. This is a project focusing on improvement of writing in the discipline areas. The project began with electronic mail/pen pal messages, but found that these were not sustainable. It soon moved to an electronic newswire service. Students contribute articles from all the sites, which include classrooms in the United States, Israel, Japan, and Mexico.

> The newswire made the connection between students more public and extended the range of types of writing and created a situation in which editing and revision were natural parts of the activities. (Riel, 1986, p. 9)

Each class may choose any of the articles on the network they wish to include in their newspaper. Writing is for a much larger audience since a student's article could be published several times. Reading is strengthened because students must carefully read each of the submitted articles in order to make their selections.

In addition to the newswire service, the participants have done cooperative projects on topics such as career aspirations, comparison of educational systems, analysis of news coverage of a particular event in different locations, and how drinking water problems are solved in each of the network locations. The science-related projects are then published in an electronic science journal (Levin, 1985).

The researchers involved with this project express sentiments similar to those of Fred D'Ignazio:

> There is a real sense in which these new electronic media can open up the world to students and teachers. Instructional networks can allow them to engage in activities far beyond the walls of their school which now so effectively separate students and teachers from the rest of society. (Levin, Riel, Miyake, & Cohen, in press, p. 10)

The Unique Aspects of Network Writing

Based on the projects just reviewed, several contributions to the teaching of writing when using a network have been identified. From the review of telecommunications projects, consequences of using a Local Area Network (LAN)—computers tied together in a limited area such as a classroom, school building, or perhaps a college campus—rather than a wide area or more global network, seem to be somewhat different. Let's first look at the results of using a LAN.

The directors of the *Quill* project concluded that the electronic mail, including a bulletin board system, was quite successful within single classes because the participants all knew one another and had a real need to communicate. In contrast, they found that the pen pals setup between people who did not know each other petered out after a while.

In the use of a LAN for science instruction, Goldman and Newman compared the procedures of a regular classroom to the one with a LAN and concluded:

> Teachers were the initiators of many interactional sequences during lessons, and more often than not, decided which directions lessons would take, summoning the students to join. In the course of face-to-face interactions, initiations and responses are crucial for the interactions to proceed. In electronic mail, this pattern was not necessary. The structure for electronic communications is much less constraining in the details of initiations and responses. Both students and teachers can initiate and both can respond. (Goldman & Newman, 1988, p. 15)

Electronic mail interactions often went over longer periods of time than a similar face-to-face interaction. This allowed several students to respond to a question, from either the teacher or a peer, which in a live discussion usually does not happen. It appeared that students had more of an equal role in thinking about science.

Geoffrey Sirc, faculty member at the University of Minnesota, taught his freshman English course using a LAN. He found the combination of interactions that were possible—student to student, faculty to student, and the whole group—to be extremely beneficial.

> Examples for discussion can be chosen and shared from students' actual work rather than from a textbook. The students' continued reading of another student's writing and the instructor's writing reinforces patterns of written English. Watching other students go from planning and pre-writing to actual composing demystifies the writing process for the most hesitant learners. Furthermore, the teacher, as just another node-name on the network, is decentralized as an authority figure. And a student with a home computer and modem can actually participate in class when unable to attend. None of these benefits is available in a traditional setting. (Sirc, 1988, p. 100)

The experiences of learning to talk about writing helped the students think about writing, and vice versa. The network provided an environment for students to *see* themselves think and see their writing and their colleagues' writing improve. The multiple paths of interaction, the "constant expression of ideas in writing paid off for my students in the quality of both their papers and their comments on the network. The saturation in writing and thinking about writing afforded by this network did much to nurture my students' sense of themselves as writers" (p. 103).

Taking advantage of Project Athena, a LAN at MIT that ties together all campus workstations, a group of faculty has experimented with ways to integrate these workstations into their writing courses. The project arose out of an interest "in applying the latest information technologies to the teaching of writing" (Lampe, 1988, p. 4). Working in the traditional classroom setting, each student now sits in front of his or her own workstation. The instructor's workstation includes a large projection screen. All classroom activities, such as handing out sample writing for discussion, commenting on students' writing, student note taking, and passing out and collecting assignments now take place over the LAN. Figure 5.1 shows the activities that are conducted electronically in class.

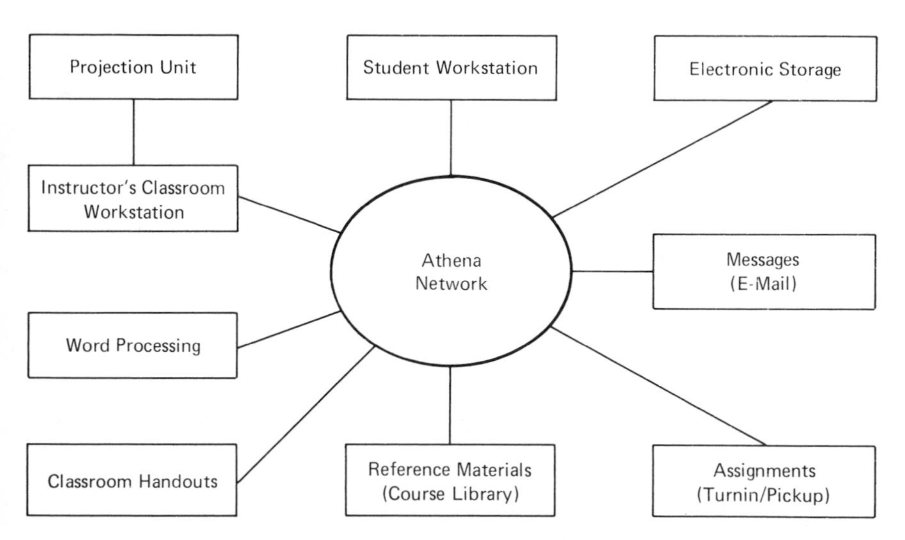

FIGURE 5.1. The Course Activities Conducted Over the Classroom LAN

When students return to their dorms, they can continue accessing all the files used in class as well as pass in and pick up homework assignments.

> So far evaluations by both students and instructors have been enthusiastic. In particular, users claimed that the ease of exchange and display of text files accelerated and enhanced the translation of theory to practice. Concepts introduced by the instructor could be tested out in the same class session, and students could get immediate feedback from their peers as well as from the instructor as they work to apply the lesson. The system made "learning by doing" a far more manageable, flexible, and useful experience than in a conventional classroom. (Lampe, 1988, pp. 5 & 13)

Goldman and Newman found the values of electronic communications within a classroom to be for:

- identified topics and group conferences
- non-competitive activities
- questions that require more reflection
- questions where answers are wanted from several students
- private interactions

They found electronic communications *not* appropriate to teach new concepts that require intense, fast-paced interactions (Goldman & Newman, 1988).

Some of the benefits attributed to the more globally networked projects such as Computer Pals and the InterCultural Learning Network, include:

- Students come into direct contact with the creators of information, the people who do the initial studies, collect the original data.
- Students have a wide audience for their writings, including peers and teachers around the world.
- They provide "a meeting place for like-minded teachers and students from all over the world . . . [to explore] cooperative educational ventures" (Waugh *et al.*, 1988, p. 5).
- Interactions, labelled "teleapprenticeships" by Levin, Kim, and Riel, "resemble those described in face-to-face apprenticeships" such as found in law and medicine (1988, p. 14).
- They provide the ability to hold tele-task forces, defined as "a group of people drawn together to accomplish some task, often for only a short term" (Levin, Kim, & Riel, 1988, p. 14).

Some of the benefits that are repeats of the benefits found with using local area networks (LANs) include:

- the need to express all ideas in writing
- use of the computer as a relatively transparent tool for academic activities
- facilitation of student collaboration
- the requirement to read carefully
- significant improvement in actual scores on standardized language tests (Bruce & Rubin, 1984; Riel, 1985; *Computer Pals Newsletter*, 1988)
- improvement in students' writing skills when writing for a real audience rather than for a term grade (D'Ignazio, 1987; Riel, 1985; Andres-Syer, 1988)
- improvement in students' self-image (D'Ignazio, 1987; Sirc, 1988)
- immediacy of translating writing theory into practice (Lampe, 1988)

BILINGUAL EDUCATION

The goals of bilingual education are essentially the same as those of English language arts—to enable students to communicate effectively in English as well as in their native language. The concern with bilingual education grows with the increasing population of Americans whose native language is not English.

Problems in Bilingual Education

In reviewing textbooks for the teaching of English as a second language, Philip Gonzales concluded that the approach taken was essentially cut-and-dried memorization: "Most materials placed great importance on 'correct' pronunciation, accurate syntax, and exact replication of language models" (1983, p. 11). Gonzales found no effort to connect the study of English to anything that relates to the real world. Furthermore, no attempt was made to integrate the different aspects of language arts—reading, writing, speaking, and listening. Finally, textbooks are set up so that divergence from the prescribed path through them is just about impossible for a teacher.

> It is suggested by this analysis that collectively publishers have for the most part decided that the way to help individuals reach the goal of language development programs need not resemble how language is normally used in real life. Real-life communication is "dynamic, interactive, and often even unpredictable". . . . On the other hand, in commercial materials, use of different and varied strategies is limited, topics are restricted, and practice opportunities are generally uniform throughout. Likewise, opportunities for creation of language for different situations, needs, and purposes is not encouraged as a rule. (p. 15)

Gonzales concludes with a plea to teachers not to use materials that are directly opposed to their intuition and logical sense of how language is used and therefore learned most successfully.

Carol Edelsky took a careful look at how, in fact, bilingual students learn to write in English. She found evidence to suggest that the teaching of language cannot be broken into small pieces; rather, it must be seen as a system of communications where all the parts are essential and are used at whatever level of development the user has at his or her command.

> What we had . . . was evidence of changes, leaps, backtracks, and appearances and disappearances of child-generated hypotheses. . . . It seems crucial, therefore, to have children be engaged with whole, authentic written discourse—to have to contend with all sub-systems at once so that they have the chance to hypothesize about something as global as audience and as local as a period. (Edelsky, 1986, p. 95)

Another pertinent area of literature helpful in explaining the successes now being experienced with telecommunications is the Modern School Movement founded by Célestin Freinet. This is a philosophy of teaching prevalent in Europe since the end of World War I. The basis of the Modern School Movement is the pairing of classes in different geographical locations for the study of a variety of topics. Information is gathered about a topic and then shared with the sister classes, initially using the postal system. Freinet claims that this correspondence between the classes:

- promotes moral development and combats egocentricism because students work cooperatively in writing their letters;
- forces the students to stop and think about their environment before answering the questions of their sister class;
- puts the students in an active learning role;
- is extremely motivating in that students have "something to say, writing to be read, to be discussed, to be responded to critically." (Sayers, 1988, p. 31)

Nothing in Freinet's model relates specifically to bilingual education. Certainly his list of benefits for two distance classes sharing information applies to any telecommunications project. However, Project Orillas, one of the successful bilingual education projects, is organized around pairing either bilingual classes or classes speaking two different languages.

Project Orillas and Other Bilingual Projects

The purpose of this project—De Orilla a Orilla (From Shore to Shore)—is to improve students' skills in writing Spanish. The project began by linking Puerto Rican students in the United States with students in Puerto Rico. Now the collaboration is extended to classes in Argentina and Canada as well. Files are exchanged at night when telephone rates are lower. Two projects currently exist: a joint newspaper and a book of Spanish proverbs which are being collected from parents and relatives.

> The most successful projects have been those which have a life of their own away from the computer, and can be amplified by the participation of the sister class. Since Orillas began a year ago, for example, student-edited newspapers have been published including articles received from "foreign correspondents." Student journalism is a particularly effective governing image because of the clear definition of roles it provides student writers; they are reporters when they write articles for local newsletters, editors while revising and polishing articles for publication, and correspondents when they send the best writings as wire-service "dispatches" to other classes around the network. Their classroom has become an international teletype. (Sayers & Brown, 1987, p. 2)

Some of the results from this project, as reported by the Office of Technology Assessment (OTA), match the results reported by D'Ignazio and Sirc:

> Anecdotal information from teachers at the sites indicates that striking results come from the English as a second language students being able to

communicate with their peers in a Spanish-language dominant society. With their newly found communicative power has come improved self images, resulting in their becoming more active participants in their regular school classes. (OTA, 1987, p. 88)

The OTA report describes another bilingual project, directed by Esteben Diaz in San Diego, which has met with success in using telecommunications:

> The network virtually allows the world to become a community resource for students in the barrio and ghetto. Students are able to "leapfrog" societal and economic barriers and create a resource network that encompasses the next neighborhood or another country. In this case, the resources provided by the network are opportunities to practice and develop literacy skills in order to communicate with their electronic friends. Friends in Spain, Harlem, or another part of San Diego are all electronically equidistant! Moreover, this means of communications operates from a presumption of equality and mutual respect that is hard to attain in face to face interactions. For students who speak another language, communications with countries in their native language reaffirms their personal heritage and underscores the value of being bilingual and literate. Students who participate in settings where access to electronic networks is part of their everyday routine developed different perspectives about themselves and the world. Communications leads to appreciation and understanding of others which leads to collaboration and cooperation in joint activities of mutual interest. (p. 122)

Because it links students in many different countries, The InterCultural Learning Network has adopted a "receiver end translation" policy to give bilingual students a chance to use their native language meaningfully in their classrooms. The following is a contribution put on the network by one of the cooperating teachers:

> I have a group of five or six students who are of Hispanic origins. While they tend to be very hard workers and extremely cooperative, a number of them struggle with their reading and writing. The articles we received from Mexico were particularly exciting. Since no one else in the class could read them the roles were changed. The articles written in Spanish illustrated to my top readers how hard it must be for the Hispanic students to read in English, and it allowed them to be in the spotlight as top notch readers. These students who have been labeled as low or poor readers, all of a sudden became the BEST readers in the class. It was a really nice experience for everybody. (Riel, 1986, p. 22)

Another teacher, Joyce Perkins from the El Paso Independent School District, reports on an activity for introducing electronic mail within her classroom. She announced that she was going to let students write notes to each other:

> Students were very receptive to the idea. . . . At least this way I would not have to intercept the note going across the room. Remember, junior high

students are much more interested in each other than any subject a computer teacher, or any teacher, could introduce. . . . Many of my students are bilingual and planned to write to each other in Spanish. (Perkins, n.d.)

English Language Education for the Hearing Impaired

English is also a second language for deaf students. As expressed by Peyton and Batson, "most have little conception of English as a conversational 'living' language, a medium of real communications, for they rarely use it for that purpose" (1986, p. 1). There are several instances reported that suggest electronic communication is an effective tool for teaching English to this population as well. In these cases, the network gives hearing impaired students opportunities to use English in both informal and formal ways. A class of deaf students is part of Project Orillas. Learning about the Spanish culture by communicating with its sister school in Rio Piedras, each class is using their second language—English.

A much more extensive experience is occurring at the Gallaudet College for hearing impaired students. Gallaudet is using local area networks in a way similar to that reported by MIT. A special room has been set up so that there is a terminal available to each member of the class. Each terminal is part of the room's LAN. The students and teacher carry on all their class conversations over the LAN. The terminals have split screens so that students are able to compose their statements on the bottom half of the screen in privacy and then push a key to send the message to the top part of the rest of the screens. The use of this network has given hearing impaired students:

- models of excellent language use in the "discussions" with their teachers;
- an opportunity for each student to develop a personality, usually done with oral language, but in this class done through written language;
- models of thinking through active, thoughtful discussions with their teachers;
- an opportunity to increase their reading ability because of their need to keep up with the discussion as it scrolls up their screens;
- increased writing abilities because writing has a real function and there is an audience—in fact several audiences: peers, teacher, and often visitors;
- increased writing abilities by watching each other and the teacher write and seeing that the process is similar for all;
- an opportunity to practice the full range of language use, from informal to formal.

FOREIGN LANGUAGE TEACHING

Foreign language courses need to teach students to communicate in a language other than their native one. Learning a foreign language is beneficial to students because it provides mental flexibility, might be useful to them later in life, or will enable them to understand other cultures better. Nevertheless, foreign language teachers have always had difficulties making their discipline meaningful to their students. Essentially, foreign language teachers have the same goals as English language and bilingual teachers: to teach their students to encode and decode in the specified language.

Background

Dewey had some ideas, put into practice in his laboratory school in Chicago, that sound very much like the Modern School Movement discussed earlier:

> "The heart of language is," Dewey said, "communications, the establishment of cooperation in an activity in which there are partners and in which the activity of each is modified and regulated by this partnership." (Herron, 1981, p. 294)

Dewey stated several conditions for the existence of meaningful communications:

1. Communications must involve partners.
2. The activity of each participant is modified by the partnership and by the situational context of the specific interaction.
3. Communication entails the sharing of experience between partners.
4. That which is communicated must be meaningful to all the participants in the interaction. (p. 294)

Dewey believed one of the major goals of foreign language study was to better understand human civilization. To accomplish this, he advised:

> . . . letters, poetry, song, drama, fiction, history, geography, engaging in rites and ceremonies hallowed by time and rich with the sense of the countless multitudes that share in them are also modes of discourse. (p. 295)

Somehow, the activities found in foreign language classes over the last 30 or so years do not always match what has been described as the activities in Dewey's school. Traditional approaches to foreign language teaching are often very teacher directed. Lessons follow the textbook closely and diverge from it only for enrichment activities. Typically, a spiral curriculum is followed, with discussions first about the neighborhood, then the town, city, and state. Speaking in the foreign language is introduced first, with reading delayed until the students have some level of spoken fluency:

> In general, reading should not be introduced until children have a good knowledge of the sound system and the most frequently used structures. When reading is begun, the initial materials should be drawn from the conversations, stories, or dialogues which children have learned or memorized. In this way, reading will constitute a visual recall of familiar spoken material. (Finochiaro, 1964, p. 28)

Writing is further delayed, with informal writing introduced first followed by more formal essay-type assignments:

> Writing should be introduced after reading. Initial writing exercises should consist of copying familiar patterns, dialogues, conversations, or reading passages. Improvisation should be discouraged until basic patterns are firmly established. (p. 28)

A few years ago, Omaggio outlined a proficiency-oriented approach to foreign language teaching. She suggests the following hypotheses and corollaries as guidelines for teachers in developing curriculum:

> Hypothesis 1. Opportunities must be provided for students to practice using language in a range of contexts likely to be encountered in the target culture.
>
> Corollary 1. Students should be encouraged to express their own meaning as early as possible after productive skills have been introduced in the course of instruction.
>
> Corollary 2. A proficiency-oriented approach promotes active communicative interaction among students.
>
> Corollary 3. Creative language practice (as opposed to exclusively manipulative or convergent practice) must be encouraged in the proficiency-oriented classroom.
>
> Corollary 4. Authentic language should be used in instruction wherever possible.
>
> Hypothesis 2. Opportunities should be provided for students to practice carrying out a range of functions likely to be necessary in dealing with others in the target culture.
>
> Hypothesis 3. There should be concern for the development of linguistic accuracy from the beginning of instruction in a proficiency-oriented approach.
>
> Hypothesis 4. Proficiency-oriented approaches should respond to the affective needs of students as well as to their cognitive needs. Students should feel motivated to learn and must be given opportunities to express their own meanings in a nonthreatening environment.
>
> Hypothesis 5. Cultural understanding must be promoted in various ways so that students are prepared to live more harmoniously in the target-language community. (Omaggio, 1986, pp. 35 & 36)

These are the ideas foreign language teachers today are attempting to implement in their classrooms. Computers and telecommunications can provide the environment created by Dewey in his school and thus aid teachers who are embracing the proficiency-oriented approach to curriculum.

Telecommunications in Foreign Language Study

On a recent visit to France, Juliette Avots, a French teacher in Wellesley High School, Wellesley, Massachusetts, was fascinated with Minitel. Distributed by the French Government and run by the telephone company, Minitel is an electronic network that ties the country together. The Minitel network was initially developed as an electronic telephone directory, but other services were soon added. Anyone can offer services on Minitel. There is a financial arrangement between the offering company and the government whereby the company retains some of the income from the sale of their services. Several educationally related services are offered through Minitel. For example, all school memos are sent through Minitel, and parents can access all sorts of school information through the network. Minitel also offers a service that matches students with tutors.

While in France, Avots had a chance to try out some of Minitel's services. Upon talking to some of the French teachers, her feeling was that they generally were not experimenting with the system because of the rigidity of the French curriculum. Nevertheless, Avots saw the potential for such a tool in her French classroom at home. Although she knew nothing about telecommunications, Avots decided to try integrating it into her curriculum. Her first project was for her students to interview, in real time, some French exchange students visiting a neighboring town. The success of this project led to several others.

The next project was an exchange with another French class in a nearby town. First the students exchanged pictures of each other. In the first electronic exchange, the students sent descriptions of each other—who they were, their unique characteristics, what they liked to do. Each class then matched the pictures with the descriptions, and the classes competed to see who could match the most correctly.

The follow-on project to the picture exchange was a round-robin story. The first class created the character and the second class the characters' personalities. The exchange continued, with classes alternately writing the actions, the crisis, and the resolution to the crisis. As each class received the story, they first corrected the mistakes of the other class. The project was originally designed as a two-week project, but actually encompassed several months. It took much longer for the students to write the story parts and the teachers to send the files than was anticipated.

Through AppleLink, Avots requested a pen pal class. Several letters were received from a class in Philadelphia, and Avots' students immediately began corresponding in French.

Avots' main problem is the difficulty in learning about an area so out of the normal flow of information in her discipline. Even though her school has excellent technical help, it all takes a great deal of time. In addition, Avots found that she and the local teachers with whom she paired for projects did not allow nearly enough face-to-face planning time, which she feels is essential. Finally, telecommunications is still an "imposition" on the curriculum. Avots recommends keeping the projects simple.

The current trend in foreign language teaching is on practical vocabulary. For the more advanced courses, more flexibility is being recommended. Wellesley High School is choosing new textbooks for next year, and Avots will be looking for ones with which she can integrate more telecommunications. She enjoyed taking her classes to the computer lab because there she did not have to direct all the activity—the students took over. She found she learned a lot more about her students by being able to wander around, look over their shoulders as they wrote, and act as a consultant. She found this teaching role to be far less threatening to the students, and as a result they were willing to write much more. The students liked using a word processor and writing for a real audience.

Avots found that the technology creates an interdisciplinary environment that helps break down the barriers people create. She found computers to be a "natural way of linking people together with information," and that they "expand ownership of information." The teacher is no longer in control of information and the student only the receptor. "Students became the initiators of their learning. They formed the questions as well as the answers. Computers integrate the world of learning and teachers become partners in learning." Avots feels that "teachers need to constantly re-evaluate their roles—what they are doing and how they are doing it. The definition of teaching is changing with the increased access to technology" (Avots, 1988).

Other ways foreign language teachers are using telecommunications include a British class studying German which assists a social studies class doing a joint research project with a class in Germany. The British class studying German translates the information sent by the German class to the British social studies class (Johnson, 1987).

A teacher of English in Japan, Hillel Weintraub, created an English language microworld by using Tom Synder's simulation, *The Other Side*.

> During the in-school practicing and online playing of the game itself, the students usually spoke English. This involved a lot of laughing, jokes, and excitement, body movements, and a lot of things which I connect to syntonic and affective learning, rather than the usual education pattern of only involving the mind. (Weintraub, 1987, p. 7)

Through this project, Weintraub feels his students truly used language

> . . . . to express feelings and ideas . . . to think in and do problem-solving with . . . to express ourselves through this self-expression, hopefully understand ourselves a little better . . . to better our understanding of others as well, and hopefully . . . to create a better world. (p. 7)

Foreign language discussions are being reported on almost all the networks used by educators—FrEdMail, the Imagination Network, the InterCultural Network, MIX. A course in Japanese is available on CompuServe's Foreign Language Education Forum, and foreign languages are being taught to the deaf on a LAN at Gallaudet.

> At beginning levels of foreign language learning, writing traditionally has been treated as peripheral, to "supplement other activities" and extended writing is often restricted to more advanced levels. At Gallaudet, foreign language classes as low as the third-semester level have been conducted successfully on the network. By the end of the semester, students in a German 201 class were participating in discussions almost entirely in German. The autobiographies they wrote in German late in the semester were longer than any the professor had ever received in previous years. (Peyton & Batson, 1986, p. 5)

CONCLUSIONS

The most striking aspect of this review of telecommunications applications to language arts is the sheer number of different kinds of applications for attaining both similar and different teaching goals. For a group of professionals who were not all that speedy to embrace word processing, they seem to have made a 180-degree turn.

Repeated Comments

Very few of the reported projects have done anything that approaches what academics would willingly label "research." However, as teachers, we know that repeated anecdotal data and gut level feelings are very important as guides to what is effective education. The integration of telecommunications in the classroom is just a continuation of the "computer revolution" that has generated enough evidence—though not necessarily research data—that educators generally agree that computers are a useful education tool. The repeating comments found in the area of telecommunications and the language arts teaching include:

- a frustrating number of problems getting started, but teachers also demonstrate "stick-to-itiveness"

- very motivating, once the bugs are worked out, when:
  - there is an acknowledged need for the tool
  - projects are of reasonable length and focused
- multiple contributions to the teaching of the language arts based on:
  - writing for a real audience rather than just for the teacher
  - writing to a more personal audience, people you know or with whom you have made acquaintance through pen pal exchanges
  - facilitating the integration of the different language arts areas as well as the other disciplines
  - encouraging students, through the use of LANs, to be more active in the educational process because the additional time allowed for more reflective answers and the role of the teacher is decentralized
  - providing hearing impaired students, through the use of LANs, an opportunity to carry on live discussions
- encouraging cooperative learning at several levels: among students, teachers, and teachers and students; by pairing classes, then students within classes
- allowing teachers to think about teaching in a more timely way with their colleagues
- the "tumbling down" of classroom walls
- a critical element for success being the role of "gatekeeper" or system manager
- starting with pen pal letters or surveys leading naturally into curricula-related projects is an effective way of introducing the tool
- increasing students' self-image by requiring skills different from traditional academic skills.

Concerns Raised

Three areas of concern have been raised by language arts teachers. First, the number of frustrations experienced in getting the system working, the lack of technical help, and the costs raise questions about the cost/benefit of telecommunications. Second, different norms of communication standards are accepted when telecommunicating. These seem to work well for students who have great difficulties with standard language practices, but could be counterproductive for students who are able to learn grammar and spelling. Teachers, however, are beginning to see telecommunications as being both formal and informal. Formal communicating, when messages are created offline, requires proper grammar and spelling. For informal, more spontaneous communicating, when messages are created online, it is acceptable occasionally to be grammatically incorrect.

Finally, the issue raised by Scott on the role of the teacher in the communication process seems worth considering. He suggests that the teacher is acting as a secretary does in an organization because of the control he or she has over what gets written. Perhaps students should have final say over what is communicated—as authors usually have final say over the content changes of a copy editor.

Usage Patterns

Classroom applications of telecommunications, based on those reported by language arts teachers, can be looked at from three perspectives: uses that imitate adult uses, uses that are being developed for attaining educational goals, and those activities that join the generations. When students take part in newspaper projects, contributing articles to a network and selecting articles for their edition from the network, they are using telecommunications as adult journalists do. When students use online databases for information for their research projects, they are using the tool as adult researchers do. When students communicate with each other to share information, such as in tele-task forces, they are imitating adults.

Such projects as the Computer Pals writing project, with its precisely delineated steps, is clearly implementing this tool from a classroom perspective. The teleapprentices concept is adapting educational experiences from other domains to the precollege classroom through the use of telecommunications. The projects that seem to be joining the generations are projects in which students conduct original data gathering and contribute to the general body of knowledge. In this chapter, the *Feeding the Billions* project is such an example. Here students are attempting to understand the plight of people in the third world by interviewing first-generation immigrants and then drawing their conclusions based on an analysis of the data. The Kids Network project, reported in Chapter 6, is of a similar ilk.

Things to Come

How will language arts teaching continue to change? What are some of the identifiable developments likely to impact the classroom? As Ronald Fitzgerald, writing for the *Christian Science Monitor* (1988) points out, research shows students learn in many different ways, but schools are ignoring this. Telecommunications, being quite effective with wide ranges of special needs students, from severely emotionally disturbed to hearing impaired, would appear to be a viable way to reach students who learn in different ways. As the tool becomes more mature and easier to use, it will provide an access route for many kinds of learners.

Digital Equipment and Houghton Mifflin have just announced a Grammar Checker, "a proofreading software package that corrects not only spelling but punctuation and syntax as well" (McGlinn, 1988, p. 7). The software uses artificial intelligence techniques. Not only does the software catch grammatical errors, it explains the error to the user. Moreover, it comments on the composition if it is either too formal or too informal. The software currently runs on a large Digital machine and will probably take some time before it is available on the garden-variety classroom computer. However, using telecommunications, it is quite conceivable students could access such software.

From the projects reported here, it seems reasonable to predict many more interdisciplinary projects, allowing the teaching of writing in the disciplines to be finally attainable. Interdisciplinary projects break down the walls within schools. Telecommunications break down the walls between schools and the rest of the world. Students will be far more easily involved in real-life problems by being able to more readily integrate all the skills and knowledge schools have traditionally found necessary to teach largely by rote or through contrived problem-solving environments.

REFERENCES

Anderson, T. (1988, February/March). CMS: Computer mail system: School Net version. *CUE Newsletter*, Vol. 10, No. 4.

Andres-Syer, Y. (1988, Spring). OCNSIDE has quite a year. *FrEdMail News*, Vol. 2, Issue 3, pp. 3, 8.

Avots, J. (1988). *Multidisciplinary team approach to technology planning*. Paper presented at Lesley College 10th Annual Computer Conference.

Baer, V. (1988, May). Getting to know the neighbors: An information exchange between two middle schools. *The Computing Teacher*, 15(8), 20–22.

Bruce, B., & Rubin, A. (1984, September). *The utilization of technology in the development of basic skills instruction: Written communications*. Report No. 6766, Bolt Beranek and Newman.

Butler, G., & Jobe, H. (1987, April). The Australian-American connection. *The Computing Teacher*, 14(7), 25–26.

Butler, Greg, Miller Computer Education Center, Miller Road, Miller NSW 2168, Australia.

Cohen, M., & Riel, M. (in press). The effect of distance audiences on students' writing. *American Educational Research Journal*.

Computer Pals Newsletter. (1988, February-April). Vol. 1, Issue 1.

Davis, G. (1987, December). Coaching writers via modem. *MacWorld*, p. 67.

D'Ignazio, F. (1987a). Setting up a multi-media classroom: A QuickStart Card. *Computer in the Schools*, 4(2), 5–29.

D'Ignazio, F. (1987b, Fall). It's the mentality, it's not the money. *Instructor*, pp. 10–12.

Dodge, J., & Dodge, B. (1987, December/January). Telecommunications. *CUE Computer Using Educators Newsletter*, Vol. 9, No. 3.

Dorey, Dennis, Jordan Small School, Box 271, Raymond, ME 04071, Telephone 202-655-4743.

Edelsky, C. (1986). *Recognizing your work in writing research: Writing a bilingual program: Habia una vez.* Norwood, NJ: Ablex.
Eliot, Richard, Lyndon Institute, Lyndon Center, VT 05850, Telephone 802-626-3357.
Finochiaro, M. (1964). *Teaching children foreign language.* New York: McGraw-Hill.
Fitzgerald, R. (1988, May 2). Reform politics overlooks the importance of learning styles. *The Christian Science Monitor,* pp. 19–20.
Goldman, S., & Newman, D. (1988). *Electronic interactions: How students and teachers organize schooling over the wires.* Paper presented at the American Educational Research Association.
Gonzales, P. (1983, Fall). An analysis of language development materials. *The Journal for the National Association for Bilingual Education, 8*(1), 5–21.
Hands on around our world. *Computer Pals Across the World,* P. O. Box 280, Manly, NSW 2096, Australia.
Herron, C. (1981). Dewey's theory of language with some implications for foreign language teaching. *Foreign Language Annals, 14*(4&5), 293–298.
Hutcher, P. (1987, September 4). MIX (McGraw-Hill Information Exchange).
Johnson, N. (1987, July). Telecommunications thrives in UK schools. *C.A.L.L. Digest,* Vol. 3, No. 5.
Lake, D. (1988). Two projects that worked: Using telecommunications as a resource in the classroom. *The Computing Teacher, 16*(4), 17–19.
Lampe, D. (1988). A new approach to computerizing the classroom. *The MIT Report, 16*(7), 4, 5, & 13.
Levin, J. (1985). *Computers as media for communication learning and development in a whole earth context.* Paper presented at the 15th Annual Symposium of the Jean Piaget Society.
Levin, J., Kim, H., & Riel, M. (1988). *An analysis of instructional electronic message interactions.* Paper presented at the American Educational Research Association.
Levin, J., Riel, M., Miyake, N., & Cohen, M. (in press). Education on the electronic frontier: Teleapprentices in globally distributed educational contexts. *Contemporary Educational Psychology.*
McGlinn, E. (1988, July 18). Software flags grammatical errors. *Boston Business Journal,* p. 7.
Mendrinos, R. (no date). *Feeding the billions: A journey through the third world.* Thurston Junior High School, 850 High Street, Westwood, MA.
Monroe, C. (1986). *Using telecommunications to ease anxiety.* Paper submitted in a graduate course in telecommunications at Lesley College.
Myers, Nina, Miller Junior High School, 6151 Rainbow Drive, San Jose, CA 95129, Telephone 408-252-3755.
NCTE. (1986). *Guidelines for the preparation of teachers of English language arts.* Urbana, IL: Author.
Oklahoma State University Teleconference. (1988, April). *Classroom Integration of Telecommunications.*
Omaggio, A. (1986). *Teaching language in context.* Boston: Heinle & Heinle.
OTA, U.S. Congress. (1987). *Trends and status of computers in schools: Use in*

Chapter I programs and use with limited English proficient students. Washington, DC: Author.

Perkins, J. RR1 Box 664R, Anthony, NM 88021.

Peyton, J., & Batson, T. (1986). Computer networking: Making connections between speech and writing. *ERC/CLL News Bulletin,* Vol. 10, No. 1, pp. 1 & 5–7.

Riel, M. (1985). The Computer Chronicles newswire: A functional learning environment for acquiring literacy skills. *Journal of Educational Computing Research, 1*(3), 317–338.

Riel, M. (1986). *The educational potential of computer networking* (Report # 16). Paper presented at the American Educational Research Association.

Rubin, A. (in press). Networking. *The Alaska Quill Network.*

Sapienza, L. (1988, March). Wayland online. *On Cue,* Vol. 1, No. 3, pp. 5, 6, & 17.

Sayers, D. (1988, April). *Interscholastic correspondence exchanges in Célestin Freinet's Modern School Movement: Implications for computer-mediated student writing networks.* Qualifying paper, Harvard Graduate School of Education.

Sayers, D., & Brown, K. (1987, July). Bilingual education, second language learning and telecommunications: A perfect fit. *C.A.L.L. Digest,* Vol. 3, No. 5.

Schrum, L., Carton, K., & Pinney, S. (1988, May). Today's tools. *The Computing Teacher, 15*(8), 31–35.

Scott, A. P. (1988, February-April). Inter-cultural communications using computer mediated communication systems. *Computer Pals Newsletter,* Vol. 1, Issue 1., pp. 11–12.

Sirc, G. (1988, April). Learning to write on a LAN. *T.H.E. Journal,* pp. 99–104.

Traversi, Dawna, Tobin School, Cambridge, MA 02138, Telephone 617-371-0202.

Walker, S. (1987). BreadNet: An on-line community. *Bread Loaf and the Schools,* Vol. 1, No. 1., pp. 12–20.

Waugh, M., Miyake, N., Levin, J., & Cohen, M. (1988). *Problem solving interactions on electronic networks.* Paper presented at the American Educational Research Association.

Weintraub, H. (1987, July). Kentucky to Kyoto, Bridge to understanding: A project in international communications. *C.A.L.L. Digest,* (3)5.

Wheeler, Fred, Behavior Management Specialist, 1682 Sullivan Drive, Norristown, PA 19401.

6
Science and Social Studies

Science and social studies focus on problem solving. Researchers in both these subject areas look for answers to real-world questions and problems. Although the questions differ widely, their solution requires information and communication. To understand the consequences of the accelerating destruction of the world's rain forests, or to find solutions to the problems of the Middle East, researchers must bring together the appropriate data and information and articulate their ideas, strategies, concepts, and values to others.

Telecommunications can greatly enhance science and social studies curricula by providing far more access to current data and information than most libraries. By expanding communications beyond our classrooms and schools, telecommunications bring together people with different perspectives and expertise. This chapter looks at how telecommunications is being used in science and social studies classrooms, and how it is changing both the types of problems students can address and how teaching and learning is accomplished in these disciplines.

TEACHING SCIENCE

Science education is in trouble. Numerous studies indicate that the science and mathematics skills of U. S. students have declined below those of other industrialized nations. The 1988 *Science Report Card,* issued by the National Assessment of Educational Progress of the Educational Testing Service, found that only 5 percent of high school students take biology, chemistry, and physics. Only 7 percent of high school students are adequately prepared for college and university science courses. Far more distressing is the fact that many students do not see the relevance of science to their everyday lives.

The problems of science education stem from a variety of factors, but the strategies to solve this crisis require not only changing the science curriculum, but also how teaching and learning occur. Plans for improving science education emphasize teaching science concepts. The focus must be on problem solving, rather than just memorizing facts. Equally important, students must perceive science as integral to their lives. To effectively develop an understanding of what science is and how to do science, students need to be able to find and use relevant information, share and discuss data and ideas, and collaborate on problem solving. Telecommunications technology is one tool available to help teachers achieve these goals.

ACCESSING INFORMATION ON SCIENCE, HEALTH, AND TECHNOLOGY

Class discussions and reports about science and science-related topics are common activities in science classrooms. For many teachers and students, getting timely information on current science topics is a problem. The reference books and textbooks available in school libraries are often out of date and therefore have little or no information on such topics as ozone depletion or the effects of the 1988 summer drought. Online commercial databases and information services such as CompuServe®, DIALOG, The Source, and Dow Jones News/Retrieval offer a wealth of up-to-date, relevant information on science, health, and technology that can be incorporated into student reports, class discussions, and other activities. (See Chapter 4 for more information on using commercial database services.)

Elementary through high school students have successfully used online databases in science classes. In Wayland, Massachusetts, high school students have researched Agent Orange, gene therapy in the treatment of phenylketonuria (PKU), and Supernova 1987A (Sapienza, 1988). Working with their teacher and school librarian, sixth-grade students in San Mateo, California, accessed *Grolier's Encyclopedia* on CompuServe to find information on the dreams of twins and on agriculture in the Soviet Union (Apple Educator News, 1986).

Students participating in Project SHINE, an interdisciplinary regional telecommunications network linking seven towns in eastern Massachusetts, combine online database searches with print materials and CD-ROM to find information on water pollution and nuclear power. Materials from these online sources were used as background information for reports and discussions on these issues with other students on the network. For example, in their study of water pollution, 150 seventh-grade students wrote letters in support of the Clean Water Act. These letters were uploaded to the WGBH science forum on Delphi and mailed to state legislators (Massachusetts Dept. of Education, 1988).

The students' research on water pollution resulted in a teleconference, *Fresh Water: How Abundant Is It?*, involving 75 students from all the Project SHINE schools. As Roxanne Mendrinos, one of the Project SHINE telecommunications team, points out, online databases were critical to the success of this activity:

> Online database searching was essential for current information. Both CompuServe and Dialog were used to access newspaper and magazine articles and up-to-date information on nuclear and water pollution issues. The card catalog, encyclopedia indexes, CD-ROM and online data bases provided students and staff with important research. We were part of a two way audio and one way video conference with a speaker from the Massachusetts Water Resources Authority who addressed seven towns in remote sites. This was a very exciting experience in distance learning. (Mendrinos, 1988)

ACCESSING WEATHER INFORMATION

In addition to helping teachers and students discuss and write about science, commercial databases and information services can help with one of the staples of the science curriculum—weather. Up-to-the minute international, national, regional, and local weather data are available through information and database services as well as specialized weather data services such as Accu-Weather and the National Geographic Weather Machine. Far more detailed than newspaper, radio, or television weather reports, the data available from these services include hourly and daily measurements of temperature, wind direction, wind speed, barometric pressure, dew point, cloud cover, fog, and visibility from hundreds of weather stations across the country and around the world.

Both Accu-Weather and the National Geographic Weather Machine have software packages to help students analyze this information. Accu-Weather Forecaster lets students display weather data downloaded from Accu-Weather as maps, tables, and graphs. The Weather Machine software displays maps of weather data at different altitudes above the earth's surface. Using such maps, students can easily follow the movement of weather patterns or see the jet stream. The manuals for both weather services contain numerous ideas about how to use weather data in classes.

In a recent article, Len Scrogan, then coordinator of computer education for the Albuquerque public schools, made several suggestions for using weather data from online services with students:

- Create some interesting, motivating scenarios that each student must research.
- Imbed the program in a larger simulation combining several other applications such as planning a presidential visit to your region (this gives the weather analysis a sense of purpose).
- Make climatic or seasonal comparisons. (Students in our classroom testing sites showed intense interest in such comparisons.) (Scrogan, 1988, p. 36)

SHARING INFORMATION

Doing science requires communication. A recent report on computer networks for scientists points out that professional scientists and engineers are relying more and more on telecommunications to share information, discuss ideas, disseminate results, and collaborate on research projects (Jennings *et al.*, 1986). Students learning science have the same need to communicate and collaborate as professional scientists. Of course, students within a science class can work collaboratively, but using telecommunications broadens their experiences. By sharing data, knowledge, and ideas with their colleagues across town or thousands of miles away, students can actually *be* scientists, building their own understanding of the world around them.

Like other professionals, teachers also need access to resources that may not be available in their schools or districts. Telecommunications can decrease the isolation of teachers, bringing them into contact with other teachers, educators, and professional scientists and engineers.

Electronic mail, bulletin boards, electronic conferences and forums, and collaborative research projects are all ways telecommunications are being used to share information and data, and support teachers and students doing science projects. These activities can be relatively small-scale, for example, on a bulletin board or electronic mail system within a single school, or international in scope.

USING TELECOMMUNICATIONS TO SUPPORT SCIENCE TEACHING

Unlike other professionals, teachers are often isolated in their classrooms. They often find it difficult to have discussions, find support among their peers, and access the resources they need to be effective teachers. Several projects in science education are developing ways to use telecommunications to break down this isolation by facilitating teacher-to-teacher interactions within a single school and among geographically dispersed schools.

Earth Lab, a recent project of the Bank Street College of Education, funded by the National Science Foundation, investigated how teachers and students could use local area networks (LANs) within their schools to collaborate and communicate about science. Working with sixth-grade classes in two New York City elementary schools, Earth Lab researchers found that the network changed how science was taught in these schools and how teachers interacted with each other. Prior to installing the network, there were few discussions between the classroom teachers and the science teacher at one of the schools involved in Earth Lab.

> The science teacher reported that he found it "difficult" to keep an ongoing conversation about science alive in the school. He pointed to several possibilities for the lack of science planning: the school schedules meant that he was teaching while the classroom teachers were on prep periods; emphasis in the school was on improving reading and writing, so most workshop or special planning times concentrated on those areas; and, quite possibly, classroom teachers had little time, interest, or training in teaching science. (Newman, et al., 1986, p. 12)

The network dramatically changed communications among the teachers:

> With the addition of the network, teachers found another way to communicate about the planning of the science curriculum. By using the network's electronic mail inside the school and hooking into the Bank Street Exchange bulletin board, teachers found a way to plan with each other without having face-to-face meetings that were difficult to arrange. Teachers received and responded to messages about science activities on the network. . . . Many of the messages dealt with coordinating events, making inquiries about activities, or sharing problems and concerns. (p. 12)

Middle and junior high school science teachers in New Jersey are using electronic communications to find resources and support among their peers. Sponsored by the New Jersey Institute of Technology (NJIT), and Fairleigh Dickinson University, teachers attend workshops every six weeks. Between workshops they use electronic mail and computer conferences on the Electronic Information Exchange System (EIES) to share curricular ideas and self-developed materials such as demonstrations, lab exercises, and activity sheets. For example, one teacher contributed this activity idea to teach students about diffusion:

> To demonstrate diffusion and the semipermeable membrane this year I had my students working in pairs: make tea with hot and cold water—there is

> quite a dramatic difference with the different temperatures, and illustrates semipermeability in a way they'll remember. (Kimmel, et al., 1987)

The goal of the New Jersey network is to give teachers the resources they need to teach science effectively. As part of this effort, teachers also use the network to find answers to their own and their students' questions from participating teachers and faculty at NJIT and Fairleigh Dickinson. One such question was, "While performing the activities in the bubble packets, a student asked why the bubbles last longer when glycerine is added to the bubbles solution" (p. 37).

Mort and Helen Sternheim set up the Physics Forum for physics teachers in Massachusetts with funding from the University of Massachusetts, Five Colleges, Inc., and the Massachusetts Corporation for Educational Telecommunications (Vaughn, 1988). Materials on the forum include test banks, lesson ideas, and public domain software. Messaging and conferences on the system let participants talk about a wide variety of subjects, from gene splicing to space exploration.

NASA Spacelink is an online database containing information on space research, NASA programs, and lesson plans and ideas for incorporating this information into science classes. The database is updated daily by NASA education specialists. Although teachers cannot send messages to each other on Spacelink, they can leave messages for NASA personnel.

Many commercial information services also have forums of interest to science teachers. Using these national services brings teachers into contact with educators and other science professionals throughout the world. Forums on CompuServe of interest to science teachers include the Educators Forums, Logo Forums, Math/Science Forums, and the Space Forum. In addition to science-related software and information databases, the Science Education Forum on CompuServe® includes:

General Science Database: Discussions for middle, junior high, and high school teachers of general science courses.

Science Update: Recent news stories about science and technology.

Job Bank: A listing of job openings in science in education and business.

COMMUNICATING ABOUT SCIENCE AND ISSUES RELATING TO SCIENCE

Electronic mail, bulletin boards, and electronic conferences also can help students think and write about science and issues relating to science. As part of Project SHINE, classes use electronic mail and video to discuss environmental issues. Middle school classes in Walpole and Westwood,

Massachusetts, are sharing reports on endangered species and environmental law through telecommunications. These classes jointly created a television show on these topics that was aired by cable stations in both towns. Other Project SHINE classes in Walpole and in Boston share information and concerns about water pollution with classes in San Francisco.

> We have received and are sending electronic mail via CompuServe to San Francisco. Our California peers are investigating San Francisco Bay while we are investigating Boston Harbor. Our students will be videotaped speaking about Boston Harbor. It is hoped the California students will videotape themselves talking about the San Francisco Bay. (Mendrinos, 1988)

Although the focus of this book is primarily precollege education, one example from undergraduate education could be readily adapted to secondary science courses. Twenty-six physics students and five professors at Boise State University, Dickinson College, New Mexico State University, Towson State University, Oberlin College, the University of West Florida, and the University of Vienna use electronic mail to solve problems as part of a unique thirteen-week physics course. Problems are mailed electronically to students twice a week. Using the network, students work collaboratively with each other and with their professors to solve the problems. The goal is to get students to come to a genuine understanding of physics by *writing* about possible solutions to a problem, not just plugging numbers into an equation. According to one student participating in the course, this approach really works.

> Einstein's ideas are very complex and sneaky and have all those lurking paradoxes. By writing them out rather than fiddling with them in your head, you come to understand them. (Marquand, 1989, p. 12)

According to Edwin Taylor, a professor participating in the course,

> The course doesn't save time. It doesn't save money or cut down the number of teachers needed. But it forces kids to clarify their thinking and communicate ideas. You want to see that. (p. 12)

For students, bulletin boards and electronic conferences can bring them in touch with people having a variety of backgrounds and perspectives. Participants in a discussion on a commercial information service can include other students, teachers, professional scientists, and interested laypersons. A high school student who needed information for a debate on AIDS posted a request in the education forum on THE WELL and received an avalanche of useful help from other participants (Apple Education News, 1986). *Science Experts On-Line* on the McGraw-Hill Information Exchange (MIX) let students ask professional scientists questions about their work. The experts included astronomers, meteorologists, and zoologists (Wigley, 1988).

COLLABORATIVE RESEARCH PROJECTS

Perhaps the most exciting use of telecommunications in science classrooms is the development of collaborative research projects. This is where student science becomes real science. Telecommunications-based projects combine hands-on activities and discovery with collaboration and sharing information. Using telecommunications, students can tackle problems that are difficult, if not impossible, to do within a single classroom, problems that require data from geographically dispersed locations or large sample sizes. The scale of telecommunications-based projects can range from those involving classes within a single school to those involving hundreds of participating schools.

FIGURE 6.1. Entering Weather Data to Share With Other Classes on the National Geographic Kids Network

Earth Lab

The Earth Lab project investigated the use of local area networks (LANs) as a mechanism for fostering collaborative investigations within a school. As part of a curriculum unit on weather, each day a group of students collected data from the school's rooftop weather station. The groups pooled their data into a common database accessible to all students through the LAN. In addition to helping communications among teachers in the school, the use of the LAN and the curricular materials also supported small group work and coordinated class investigations (Newman *et al.*, 1988).

> While the system was a tool for teachers, it was also a management tool for students which removed some of the burden from the teachers and which helped the students to take more control of group work themselves. (p. 9)

An example of the power of this teaching technique is illustrated by the report of how students "discovered" the theory of continental drift:

> One large piece of evidence supporting the theory of continental drift is the discrepancy between the climate of the locations where fossil remains have been found and the type of climate that the fossils could actually live in. In order to model how real-world scientists used fossil evidence to discover the theory of continental drift, we created a fossil database for students to manipulate. (Brienne & Goldman, 1988, pp. 6–7)

The students then created two maps, showing where each fossil was found and then where it could have lived based on climate considerations.

> Students began to see the discrepant information that the two maps yielded. For example, they observed that the New York State fossil, Eurypterids . . . could only have lived in a tropical climate. And that Eolian Sand (sand dunes) found in England really belonged in a tropical dry climate. They began to generate theories about how these fossils ended up in hostile climates. Students were eventually led to suggest the theory of continental drift. (pp. 7–8)

WaterNet

Developing a shared database of information is also part of WaterNet, a University of Michigan project funded by the Department of Education. Using a curriculum developed by science, computer education, and social studies teachers, WaterNet combined telecommunications with microcomputer-based laboratory tools (MBL), databases, spreadsheets, graphing packages, and interactive computer simulations. According to Carl Berger and Clancy Wolf:

> The purpose of this two phase project was threefold: 1) to develop the skills necessary to solve problems of important local, social, and environmental issues, 2) to develop a critically aware and responsive citizenry with the motivation and desire to act upon social-ecological issues, and 3) to establish a regional network for the achievement of these goals. (Berger & Wolf, 1988)

High school students from the United States and Germany gathered water quality data from their local rivers using commercial water pollution kits to measure temperature, turbidity, pH, dissolved oxygen, fecal coliform, biological oxygen demand, total phosphorus, total nitrates, and total solids. They also used microcomputers connected to simple analog-to-digital conversion devices to automate continual monitoring and collect timed samples. Data were collected from the South Platte in Colorado, the Duwamish in Washington, the Rouge River in Michigan, and the Rhein River in West Germany. Then, using CONFER II, a conferencing system on the University of Michigan's mainframe computer, students pooled their data into a shared database accessible by all schools. Working with this database, students used the electronic mail and conferencing facilities on CONFER to question each other about the data from different river systems, to discuss the implications of the results, and to suggest new approaches.

FIGURE 6.2. Collecting Water for Analysis

> By sharing the results of the tests that each school conducts, students are able to see a wide variety of values for each of the tests. The ability to communicate easily and inexpensively with other locations involved allows for a rich hypothesis generating environment and the ability to quickly run follow-up explorations.

Berger and Wolf point out that this project was not without its problems. The majority of difficulties involved getting access to computers, modems, and a phone line, or problems using the CONFER II teleconference system. In implementing collaborative projects, they suggest allowing adequate startup time to overcome technical and logistic problems. Berger and Wolf also point out that the WaterNet's success is because teachers and students had a reason to communicate—without a purpose, communications are not sustained.

The Intercultural Learning Network

Although primarily focusing on social studies and language arts, the Intercultural Learning Network, a project of the School of Education, University of California, San Diego, also developed several telecommunications-based science projects. These science projects included both observational activities as well as collaborative problem-solving activities (Levin & Cohen, 1985; Waugh *et al.*, 1988).

In the Boiling Hot Project, students from the United States, Japan, Israel, and Mexico studied how water boils. First, students at different locations wrote down their expectations of the temperature at which water would boil and shared these on the network. They then measured the actual temperature at which water did boil and shared the data. After receiving the results from other schools, discussions began on the network about other students' procedures and experimental errors, for example, is a measurement of 99.9°C the same as 100.1°C? With data from a number of sites, students ultimately developed an understanding of how elevation affects the temperature at which water boils.

Integrating science, social studies, and language arts, the Water Problem Solving Project is an example of a collaborative research project on the Intercultural Learning Network. In 1985 and 1986, students from schools in the United States, Mexico, Japan, and Israel studied the real-world problem of shortages of drinking water by collecting data on their own community's water supply (Levin & Cohen, 1985; Levin, *et al.*, 1987; Waugh, *et al.*, 1988). After analyzing the similarities and differences in how different communities obtain water, students looked at the feasibility of adapting solutions used in other parts of the world to solving their own local water problems. Other problem-solving projects on the network include studies of severe weather conditions, animal and insect pests, pollution, and energy. In selecting projects, Waugh and Levin (1989) suggest that they should:

1. integrate well with ongoing science activities in the classroom;
2. draw upon phenomena which are common across locations, yet vary significantly based on geography, geology, ecology, culture, etc;
3. involve phenomena where the observed differences will be large enough to emerge through the potentially large errors of measurement;
4. require measurements which are fairly simple to make;

5. involve data which are not otherwise easily obtainable;
6. possess component activities (discussions, etc.) which can be conducted in the absence of network activity (can proceed while waiting for network responses);
7. involve students in all stages (data collection, interpretation and report writing activities), not just as collectors of data to be analyzed by remote adult experts. (p. 30)

Students on the Intercultural Learning Network also shared written reports of their work through the *TeleScience Chronicles*, an electronic science journal. In fact, a project was not considered complete until a written report was published in the *Chronicles* (Waugh *et al.*, 1988). Interestingly, in a related study on students' writing on the network, Cohen and Riel (1986) found that students' writing was far better when they wrote to their peers at other schools than when they wrote compositions on similar topics for actual grades.

MIX

Through the Plant Growing Contest on MIX, elementary students in the United States and Canada used telecommunications to develop an understanding of how to design a scientific experiment (Wigley, 1988; Schrum, *et al.*, 1988). On the same day, all of the classes planted corn seeds supplied by MIX. Then, using the network, they sent measurements of the height of the corn to the other participating schools every two or three days. Students in each class experimented with water, light, and fertilizer, attempting to grow the tallest corn plant. In addition, using the network, students were able to ask professional scientists about their strategies and share their results with them.

The MIX contest motivates students to become involved in science. According to Jeff Holte, one of the coordinators of the contest and a teacher of a class who participated in the contest:

> . . . the interaction made the experiment real. . . . The students learned things they will always remember because they did true comparisons of real live information. (Solomon, 1989, p. 22)

NORSTAR

The Norfolk Technical Vocational Center, a vocational high school in Norfolk, Virginia, is also a NASA research laboratory (NORSTAR). Students at Norfolk Tech are developing experiments for a 1990 space shuttle flight. Connected through telecommunications to schools participating in similar projects in the United States, Canada, Norway, and France, students discuss their ideas with their fellow researchers. In addition, using the network, they and their teachers can get assistance from

NASA personnel. Ron Fortunato, director of the NORSTAR lab, comments that this experience has redefined his role in the classroom:

> My role as teacher has changed. I'm a mentor to the students, but the lab is part of the real world. The students are engaged in life-long learning skills, creative thinking, problem-solving, communication. They make mistakes. My job is to direct them to the resources that will help them to recognize their mistakes and hopefully to avoid them in the future. Of course, they're not the only ones learning. We're all learning, the students, myself, and the NASA team that works with us. This is fun. ("Restructuring and technology," 1988, p. 4)

The National Geographic Kids Network

With funding from the National Science Foundation and the National Geographic Society, the Technical Education Research Centers (TERC) are developing the National Geographic Kids Network. This telecommunications-based science curriculum for fourth- through sixth-grade students currently involves about 600 schools in the United States, Canada, Israel, Argentina, and Japan.

FIGURE 6.3. Analyzing Data from the Kids Network

Each six-week Kids Network unit focuses on a different current science topic, such as acid rain, weather, and water quality. Through a series of hands-on activities, students collect data by measuring the acidity of their local rain or the lead content of their school's tap water, for example. They then share these data with other schools on the network. The Kids Network software lets students display their own data and the combined results from all schools as tables, graphs, or maps. Using these multiple representations, they can look for patterns in the data and compare their own findings with the national picture.

Students also use electronic mail to share ideas, the results of their experiments, and discuss what the data mean.

> Our names are Jamie and Stefanie. We are doing papers about our observations on living and non-living things. We did one experiment on living things. We took a coffee filter and put 5 mung seeds in each of 3 cups. We put pH 2.2 in one cup pH 4.0 in another cup and we put pH 6.0 in the third cup. Of course the seeds only worked in pH 6.0 which was tap water. The seeds couldn't work in pH 2.2. Neither could they work in pH 4.0 cause it was too acidic. Write back if you have any questions.
>
> Your friends,
> Jamie and Stephanie

> Dear Leramie [sic],
>
> Here are some questions I want to ask you: How big is your town, how many people are in your class, how many kids in your class go to school by bus, how many walk to school, and how does the teacher get to school?
>
> Signed,
> Eric

> Dear Team Members,
>
> Our water is unsafe! We found that we have 45 ppb of lead in our water. We also found that the pH of our water is 6.0. We also had another test by a lab in our city. The water was taken from the same sample that we used to get 45 ppb. The result was 48 ppb. We also had a microbiologist test our water for pH. He used a special pH meter to test our water and got a pH of 6.75 . . . Here are our solutions:
>
> 1) We should test the water again by flushing the pipes and retesting for lead.
> 2) If that doesn't work, the city should raise the pH of the water at the Water Treatment plant to cut back on the corrosion of the water pipes leading into the school and in the school.
> 3) If that doesn't work, the school should put some kind of devices on the faucets to take out the lead.
>
> We are concerned about the lead in our water and worried about the health of the younger students attending Riverview School. WE FEEL SOMETHING MUST BE DONE TO CORRECT THIS PROBLEM IMMEDIATELY!! We

learned about health problems that can result from lead poisoning and we're all very concerned.

Sincerely,
Riverview Sixth Grade Scientists
Wausau WI (TERC, 1988)

A professional scientist doing research in the area of the unit works with the students. Using the network, the unit scientist sends electronic mail letters to schools, commenting on the students' data, suggesting ways to look at their data, and giving teachers and students additional information about the topic. For example, in the acid rain unit, students had questions about scrubbers. Through electronic mail, the unit scientist sent the information to the schools:

> Scrubbers are devices used in power plants and factories to remove pollutants before they come out of the smoke stack. There are several kinds. Almost all plants now have scrubbers that remove the soot particles so only the gases escape. Now there are scrubbers that can remove a part of the sulfur and nitrogen gases that cause acid rain. However not all plants have these and they are very expensive. (TERC, 1988)

By combining hands-on activities, cooperative inquiry, meaningful scientific problems, and telecommunications, the goal of the Kids Network is to have students do science:

> Through the Kids Network curriculum activities, students learn about specific content areas (e.g., geography or acid rain); however, the emphasis is on having students do science by asking good questions, collecting both quantitative and qualitative information, and learning to use this information to answer significant scientific questions. (Foster, *et al.*, 1988, p. 38)

The TERC Star Schools Project

The TERC Star Schools Project is creating a telecommunications-based science and mathematics curriculum for middle and high school students. Funded by the Star Schools Program of the U. S. Department of Education, the goal of the TERC project is to involve students in cooperative investigations in science and mathematics. Each unit focuses on a specific problem. The first five units cover radon, weather, iteration (a mathematics unit), polling, and design. In each unit, students first undertake a core activity and share the results of this activity with other participating schools. For example, in the radon unit, students measure the radon levels in schools and homes in their region. After completing the core activity, students are encouraged to design their own research projects with other students on the network. Classes in Colorado, New York, and Minnesota might decide to work together on the biological effects of radon.

The network enables classes to find others interested in working on the same investigations.

To help teachers to effectively incorporate technology and project-based mathematics and science into their classes, TERC and ten Resource Centers are providing extensive teacher training and support through workshops and telecommunications. The Resource Centers include Arizona State University, Boston Museum of Science, The City College of CUNY, Minnesota Educational Computing Corporation (MECC), Northwest Regional Laboratories, Pepperdine University, Tufts University, University of Michigan, and the University of Virginia.

TEACHING SOCIAL STUDIES

Social studies also requires information and communication. Online databases, bulletin boards, forums, electronic mail, and collaborative projects can all be used to enhance the social studies curriculum. The ability to go outside the classroom to discuss issues and share ideas with people from different backgrounds and perspectives makes using telecommunications in the social studies classroom extremely powerful.

USING DATABASES AND INFORMATION SERVICES

Using CompuServe®, Dow Jones New/Retrieval, Einstein, and Western Union, students can find articles from national and regional newspapers. Robert Vlahakis, a teacher at Shoreham-Wading River Middle School, Shoreham, New York, and his students used these services to compare how news is reported in different areas. They selected eleven newspapers located in different regions throughout the United States. They also included TASS, the Soviet news agency, in their study. The topic chosen was the then-upcoming Gorbachev-Reagan summit in Washington, D.C. Using Einstein, students searched for and collected articles over several days. With so much information to collect, Vlahakis had to reorganize his class.

> . . . . we quickly had to develop systems for organizing, storing, editing, and printing articles. I tried to simplify the task to fit into the confines of several 40-minute classes. Since this was my seventh graders' first experience with Einstein, I decided to designate several students to become our "telecommunications experts," taking responsibility for gathering the articles. . . . After experimenting with the steps involved in searching for newspaper articles online, we decided that the most efficient approach was to collect and store two or three articles in one class period and then format, edit, and print on another day. (Vlahakis, 1988, p. 83)

To analyze the articles, students grouped them by geographical locations and also by time (before, during, and after the summit). Students found that, contrary to their ideas, the Soviet articles were very positive about the summit. Although working with a small data set, students did find differences in how newspapers covered the summit, some taking a negative stance, others strongly positive. Vlahakis suggests other topics suitable for "newspaper analysis," such as AIDS, apartheid, world hunger, and waste disposal. His suggestions for incorporating this type of telecommunications-based activity in a class include assigning groups of students to collect and summarize articles from a particular newspaper for the class, look for national trends as well as regional differences, get the parents involved by having them read articles and discuss them with their children, conduct polls on the topic before and after the activity, and have students write their own articles on the topic.

SHARING IDEAS AND INFORMATION

Bulletin boards, electronic conferences and forums, and electronic mail are all tools that can be used to promote discussions and sharing information. At San Leandro High School, San Leandro, California, students run the "Open Campus" bulletin board (*Apple Education News,* 1986). Forums of interest to social studies teachers on this school-based system include the Jolly Roger, where students and teachers post messages and comments concerning the student government, and The Chestnut Tree Cafe, named for the cafe where dissidents met in Orwell's *1984*. In the Chestnut Tree Cafe, teachers post opinions and comments about current social and political issues. Through the bulletin board system, student and teachers then debate these issues. Similarly, on the Schoolnet Forum, students at Project SHINE schools in Westwood and Brookline, Massachusetts, discuss alcohol and drug abuse, smoking, and peer pressure (Mendrinos, 1988). Other Project SHINE schools have used telecommunications to discuss nuclear power and nuclear war. The nuclear power forum included the following questions and responses:

> *In society, which rights predominate—the individual's right to safety or the society's need for power?*
>
> I think the people's safety should come first.
>
> I think both are important. Safety is more important, but I'm not sure we could all live without power.
>
> *What kinds of messages do the incidents at Brown's Ferry, Three Mile Island, and Chernobyl send to you?*

> These are only three accidents of the many running power plants today. These accidents would not stop me from encouraging the use of nuclear power plants.
>
> They send messages like, people are careless and don't think. It shows me that power plants aren't safe and that plants aren't tested enough. (Massachusetts Dept. of Education, 1988, p. 20)

Students can learn more about other regions and cultures through electronic mail. Schools in one part of the United States have developed telecommunications-based letter writing projects with schools in other parts of the United States and other countries. Telecommunications link gifted and talented students at middle schools on the Eastern Shore of Maryland with schools in the western part of the state. Vicki Baer, one of the originators of the projects, describes the activities:

> . . . . students were paired according to their desire to share information about geographic characteristics, historical events, historical figures, historical landmarks, industries, careers, or community activities unique to their counties. . . . Eager to share facts about their home counties, the students spent many hours scouring reference materials, interviewing county residents, and compiling reports. (Baer, 1988, p. 21)

The Maryland project involved collaborations between the social studies, language arts, and pull-out program teachers and the computer coordinator at each school.

> The social studies teacher and the pull-out program teacher provided class time for information gathering, computer lab time for composing and editing and general guidelines for the final reports' format. The computer lab coordinators provided technical assistance for word processing, for appending the final reports into one large file for transfer and the actual file transfer via modem. (p. 21)

Students exchanged four letters throughout the school year. The highlight of the project was a visit by the Eastern Shore students to their western Maryland friends. The Washington County (western Maryland) Chamber of Commerce created a handbook of the students' reports for distribution to fifth-grade social studies classes in the county. Baer reports that because students were communicating to their peers, they felt very responsible for the accuracy of the information they included in their writing.

Schools are also communicating with classes in other countries, sharing cultural information. Alan November's classes in Wellesley, Massachusetts, correspond with schools in Great Britain. Greg Butler, a teacher at the Miller Computer Education Center, Miller, New South Wales, Australia, describes electronic mail communications between his fifth-year students and American children from an elementary school in West Pottsgrove, Pennsylvania.

> Many discussions developed over the different culture, school organization, TV programs, toys, sports, and games. The many similarities between the two countries also provided the opportunity for discussion. The opportunities to look at maps, research books and to consider how America is similar and different were available. (Butler, 1988)

In addition to letter writing, students are also using electronic mail as newswire services. In many cases, the stories in these electronically generated newsletters include articles about history, culture, and geography. A recent edition of the MIX Student NewsWire contained stories from students in Australia, Canada, and the United States. Each story was written by one or more students and then sent via MIX to the editor, also a teacher. In addition to poems and fiction, articles in the resulting newsletter also discussed Australia's Bicentennial celebrations and a visit to a Soviet summer camp.

Involving about 350 elementary and secondary schools in the United States, West Germany, Holland, France, Australia, Canada, and Japan, the Long-Distance Learning Network (LDLN) combines geography, social studies, language arts and science activities (Riel, 1988). Classrooms are grouped into "Learning Circles," each circle consisting of approximately 10 classes. There are two types of Learning Circles on the network: Publishing Learning Circles, which participate in collaborative writing projects including journals and newspapers, and Researching Learning Circles, which conduct cooperative research projects in geography, social sciences, and science. Although Researching Learning Circles focus on collaborative projects, the final goal of each project is to achieve a written report of the group's work. Here's the electronic mail letter that started the Compchron: Two (the group's name) Learning Circle newsletter:

> Here's what you have to do: Organize your students into small groups who would like to work on one or more of the following topics:
>
> For ten days, have the group gather information about:
> A local news story
> An international news story
> A local sports story
> A national, international sports story
> A local interest/entertainment type feature
> An international interest/entertainment type feature
>
> After students have gathered material they should write some sort of feature article about the topic chosen. Have your student material sent in to me by Tuesday, April 26th. (AT&T Long Distance Learning Network, 1988)

COLLABORATIVE PROJECTS

Collaborative projects involving data collection, analysis, and discussion can involve students within a single school or schools throughout the world. In the Buffalo, New York, area, 2100 students at fourteen schools voted electronically. Then, using database and spreadsheet programs, the students compared the results of their vote to the voting patterns of adults in their area (*Apple Education News*, 1986).

Opinion polls are an important component of decision making in this country. At the Region 4 Teacher Education and Computer Center, California State University, Sacramento, California, Martin Harris and his colleagues (1987) are creating a series of study units for seventh- through twelfth-grade government and civics classes to help students understand polls. Using telecommunications, students will share their opinions on current issues and analyze the results of the polls. The TERC Star Schools Project is also developing a unit on polling. Designed for middle and high school social studies, science, and mathematics classes, the unit allows students to create and analyze their own polls.

The topic of one student poll was: Are manned space explorations beyond the Earth's orbit necessary? With the help of CompuServe®'s Space Forum administrator, three engineering students from Worcester Polytechnic Institute, Worcester, Massachusetts, used CompuServe®'s electronic mail system to poll 2400 CompuServe® users (Conroy, 1988). The survey found that most of the users surveyed supported manned spaceflights, but not strictly on economic grounds; humanistic and intangible reasons were just as important.

Roxanne Mendrinos and her students in Westwood, Massachusetts, are developing a database on recent immigration to this country. They sent the following survey to schools in Massachusetts and California:

1. Your place of birth.
2. What country did your parents come from?
3. When did they leave?
4. What years of school did your parents complete and what did they study?
5. What special skills do your parents have?
6. What jobs have they had since arriving?
7. What are their jobs now?
8. Describe your home. Is it a single family house, apartment, or condo?
9. What difficulties have they had in adjusting to the new country?
10. What are their likes and dislikes regarding living here?
11. List the town you and they live in. (Mendrinos, 1988)

One school in California which responded to the Westwood students' survey had 273 students who spoke 23 different languages. Schools in the

Boston area also studied immigration, writing and sharing accounts of the experiences of Irish immigrants based on historical data.

The Places One Learning Circle's description of their activities gives a good idea of the ways schools are using the Long-Distance Learning Network to develop collaborative projects:

> Participation in our Places One Learning Circle has included students in grades 4–8. In the beginning several objectives were set up. They were to compare and contrast the history and geography of the respective places represented in our group. To achieve these goals some of the specific activities we have developed and are actively pursuing include:
>
> 1. Discovering, retelling, and exchanging of local legends.
>   a. Australian (Aboriginal, animal)
>   b. Indian (Native American)
>   c. State (New Jersey, Indiana, California, New York)
>   d. Dutch
> 2. Research and documentation of local history and geography.
>   a. Interviewing
>   b. Surveying of students
>   c. Creating databases
>   d. Writing chronicles
>   e. Creating a simulated bicycle trek.
>
> (AT&T Long Distance Learning Network, 1988)

INTERACTIVE COMPUTER SIMULATIONS

Telecommunications-based simulations are being used to help students understand current world issues. Although it can be easily used within a single classroom, students in two classes at different locations can participate in Tom Synder's *The Other Side* through telecommunications. Using *The Other Side*, students attempt to build a bridge between two countries. To accomplish this seemingly simple task, they quickly find they must negotiate with each other to maintain a stable economy and national security.

The Interactive Computer Simulations (ICS), developed by the University of Michigan School of Education, bring students together through telecommunications to study public affairs. The two ICS exercises currently available are about the Arab-Israeli conflict and the American Constitution.

ICS is a role-playing exercise. Guided by the University of Michigan-based Control Group and their school facilitator, students work through a series of research activities prior to the simulation. Their goal is to become the political leaders they represent, understanding their politics and personalities. These preparation activities last about seven weeks and involve daily electronic communications to read and comment on informational items sent from the Control Group.

The actual simulation, done through the CONFER II computer conferencing system, takes about a month. Students are presented with a "scenario." They then use the telecommunications system to solve the conflict. Interactions include private diplomatic messages, press releases, political, military, and economic actions, and meetings—all through the network.

Debriefing sessions follow the simulation. Students discuss their approaches to the problem and how different strategies worked or didn't work. The purpose of the debriefing is to help students organize their learning and develop an understanding of what they learned about the specific issues of the simulation and national and international politics.

HOW TELECOMMUNICATIONS ENHANCES SCIENCE AND SOCIAL STUDIES EDUCATION

Several common themes emerge from this survey of some of the ways teachers and students are using telecommunications to enhance science and social studies classes. First, telecommunications *expands the amount of information* available to teachers and students. Whether students are searching an online database for information on corporate takeovers, planning a research project on groundwater problems with students in Australia, or conducting a poll on students' opinions about science, they can easily get the information they need using telecommunications networks.

Second, telecommunications *promotes collaboration and communication* between individuals, within a class, or among classes on a network. Students using database services must work closely with librarians and teachers to plan their search strategies. A class preparing pH measurements to send to another class must collectively decide how to gather the information and how to best organize it so that others will understand it. Classes from California and Israel working together on a water resources project must first collaborate in developing an understanding of the similarities and differences between their communities so that they can work together to solve the problem of getting adequate water supplies.

Third, telecommunications promotes an *interdisciplinary approach* to science. In these telecommunications projects, language arts, geography, social studies, and science are integrated in relevant and meaningful ways. When sharing information with classes in Millstadt, Illinois, and Noatak, Alaska, it becomes important to know where those places are and what they are like.

Finally, telecommunications *expands the boundaries of the classroom*. Using telecommunications networks, students can explore problems, such as immigration and acid rain, that are more relevant and meaningful when studied by geographically dispersed classes. In addition, students must work with people with different opinions, giving them the opportunity to view problems from a variety of perspectives.

In an editorial on the potential for telecommunications networks to change education, Jason Ohler (1987) points out that the complexity of our world and vast amounts of information needed to find solutions to problems make it crucial that students learn to work together effectively. Telecommunications networks not only give students access to the information they need to make informed decisions, they also give students the ability to communicate and collaborate as easily with their colleagues in Mexico as they do with the class across the hall.

REFERENCES

Apple Education News. (1986, January-March). Apple Computer, Inc., Cupertino, CA.

AT&T Long Distance Learning Network. (1988, April). Newsletter.

Baer, V. E. (1988, May). Getting to know the neighbors: An information exchange between two middle schools. *The Computing Teacher, 15*(8), 20–22.

Berger, C. F., & Wolf, C. J. (1988). *Using technology to manage complexity: Water monitoring in local streams and rivers.* Paper presented at the International Association for Computing in Education.

Brienne, D., & Goldman, S. (1988). *Collaborative network activities for elementary earth science.* Working Paper. New York: Center for Children and Technology, Bank Street College of Education.

Butler, G. (1988). Personal communication.

Cohen, M., & Riel, M. (1986). *Computer networks: Creating real audiences for students' writing.* Technical Report No. 15. San Diego, CA: Interactive Technology Laboratory, University of California.

Conroy, C. (1988). Outer limits. *Online Today,* p. 7.

Foster, J., Julyan, C. L., & Mokros, J. (1988). The National Geographic Kids Network from Technical Education Research Centers, Inc. (TERC). *Science and Children, 25*(8), 38.

Harris, M., Hannah, L., Matus, C., & Watson, W., Jr. (1987, April). The California Student Opinion Poll Project. *The Computing Teacher, 14*(7), 33.

Jennings, D. M., Landwebber, L. H., Fuchs, I. H., Farber, D. J., & Adrion, W. R. (1986). Computer networking for scientists. *Science, 231*, 943–950.

Kimmel, H., Kerr, E., & O'Shea, M. (1987, November). Computerized collaboration: Taking teachers out of isolation. *Computing Teacher,* 36–38.

Levin, J. A., & Cohen, M. (1985). The world as an international science laboratory: Electronic networks for science instruction and problem solving. *Journal of Computers in Mathematics and Science Teaching, 4*(1), 33–35.

Levin, J. A., Riel, M., Miyake, N., and Cohen, M. (1987). Education on the electronic frontier: Teleapprentices in globally distributed educational contexts. *Contemporary Educational Psychology, 12*, 254–260.

Marquand, R. (1989, February 17). Physics free-for-all via computer. *The Christian Science Monitor*, p. 12.

Massachusetts Department of Education. (1988). *Project SHINE: Sharing human and information networks in education*. Boston, MA: Author.

Mendrinos, R. (1988). *Educational technology and the curriculum: An interdisciplinary approach to learning*. Edmund W. Thurston Junior High School, Westwood, MA.

National Assessment of Educational Progress. (1988). *The science report card*. Princeton, NJ: Educational Testing Service.

Newman, D., Goldman, S. V., Brienne, D., & Jackson, I. (1986, May 1-December 31). "*Earth Lab: Progress report*. New York: Center for Children and Technology, Bank Street College of Education.

Newman, D., Brienne, D., Goldman, S., Jackson, I., & Magzamen, S. (1988, April). *Computer mediation of collaborative science investigations*. Revision of paper presented at the symposium on Socializing Children into Science, American Educational Research Association.

Ohler, J. (1987, November-December). Laptops, networks, and the evolution of education. *Electronic Learning, 7*(3), 8–10.

Restructuring and technology: Part one. (1988, October-November). *Radius*, a publication of the AFT Center for Restructuring, Vol. 1, No. 3, p. 4.

Riel, M. (1988). Personal communication.

Sapienza, L. (1988, March). Wayland online. *On Cue*, Vol. 1, No. 3, pp. 5, 6, & 17.

Schrum, L., Carton, K., & Pinney, S. (1988, May). Today's tools. *The Computing Teacher, 15*(8), 31, 34–35.

Scrogan, L. (1988, September). Software reviews: Accu-Weather Forecaster. *Classroom Computer Learning, 9*(1), 34–36.

Solomon, G. (1989). Hands-on science projects with help from online networks. *Electronic Learning, 8*(6), 22–23.

TERC (Technical Education Research Centers). (1988). Personal communication.

Vlahakis, R. (1988, May-June). From TASS to Tallahassee: In search of today's news. *Classroom Computer Learning*, pp. 82–87.

Vaughn, T. (1988, March). Don't be isolated: Telecommunicate. *On Cue*, Vol. 1, No. 3, pp. 11, 14.

Waugh, M., & Levin, J. (1989). TeleScience activities: Educational uses of electronic networks. *The Journal of Computers in Mathematics and Science Teaching, 8*(2), 29–33.

Waugh, M., Miyake, N., Levin, J., & Cohen, M. (1988, April). *Problem solving interactions on electronic networks*. Paper presented at the American Educational Research Association meetings, New Orleans, LA.

Wigley, G. (1988). Personal communication.

7
Managing Telecommunications

How do telecommunications programs actually get implemented in a real school? Where does the money come from? What are the steps usually taken to get started? What are the critical elements? Where are the pitfalls? This chapter addresses these questions.

Establishing telecommunications at the school level involves more than the purchase of hardware and software and access to a phone line. Successful, long-term integration of telecommunications into the learning environment of any school requires vision, support, and planning by many individuals. It may well require a phase-in period of several years to develop the curricular relevancy, budgetary allocation, and hardware and software system. Starting small and planning for growth is an excellent way to explore the educational potential of this emerging technology.

As with many other fields, new methodologies in education often encounter resistance. Change is unsettling, and those who feel responsible for the educational climate of the school, but who do not actually teach—principals, supervisors, and parents—often are reluctant to experiment with new ideas. Their support, however, is critical for both philosophical and practical reasons. Philosophically, they are an active part of the learning environment. These persons need to believe that the uses to which telecommunications are put represent an improvement in children's education. Practically, they control the purse strings and, like any other aspect of the school curriculum, telecommunications programs have budgetary requirements. Winning the support of these persons is worth the time and effort required. It is a two-step process: laying the groundwork and developing a strategy, and requires doing homework and becoming knowledgeable. Enlisting the support of others requires responsible arguments, confident answers to hard questions, and actual demonstrations of the educational potential of telecommunications.

PREPARING THE GROUNDWORK

If telecommunications is to further educational efforts, it must be seen as an effective means of meeting the school's philosophy, operative rationales, and curricular goals. These superstructures are explicitly espoused by most schools, and tying telecommunications to them in specific terms is an effective way to make a rational case for the inclusion of this technology. Most schools' statements of philosophy acknowledge the school as an active agent in preparing children to become constructive members of society. Telecommunications technology is rapidly becoming an important part of our society. It is therefore the challenge and responsibility of every school to integrate telecommunications into its curriculum.

Acceptance of the importance of telecommunications is not the same thing as giving it priority in the school. To attain the attention it deserves, good reasons for including telecommunications must be put forth. Two important facets are its "realness" and its fundamental nature. Telecommunications is real; it is the way the world works now. It will only become more pervasive. At present, it is the skeleton supporting much of the world's communication, transportation, and commerce. Students already live in a world with international direct-dial telephone service, live satellite sports and news broadcasts, automated banking, networked store checkout registers, and public libraries with online research services. Many of these services can be used through home computers. Telecommunications is as relevant and basic today as reading, writing, and 'rithmetic are.

Second, telecommunications has changed the purpose of the telephone. No longer is the telephone just a business tool or personal diversion, it is now a window onto the world of information. Understanding this is key to supporting telecommunications in the school. Specific curricular goals that demonstrate this rationale help convince principals, curriculum directors, and parents of the significance of telecommunications.

Three established curricular goals adapt well to telecommunications: communication skills, collaborative skills, and thinking skills. Telecommunications can foster communication skills in both the interpersonal and technical sense. From an interpersonal perspective, it offers social contact with a wide circle of people, real audiences for written work, and multicultural interaction on an international scope. In a technical sense, both the nature of effective communication and the underlying electronic principles can become topics of study.

Collaborative skills and group or cooperative learning are facilitated in many ways by telecommunications. Group sharing, both of resources and results, yields positive social growth. Cooperative planning and teamwork can be objectives specifically included in projects and activities, and effectively monitored and assessed through student performance. Collective decision making draws upon the advantages of teleconferencing and real-time access to information. Telecommunications encourages active participation and rewards with rapid feedback.

Thinking is a most necessary skill, especially at a time when information is so crucial. The ability to search for, retrieve, process, analyze, and interpret data defines "literacy" now. Students must practice these skills if they are to develop proficiency, and telecommunications provides the means. Practicing telecommunications is as important as learning to read and write. In fact, it can be used to help students learn to read and write. Just as with paper, pencil, and books in the past, access to telecommunications has become another requirement for learning "the basics."

Careful consideration of why telecommunications is important in the educational life of students goes a long way toward building a supportive climate in the school. However, for many people, the difference between just thinking it has possibilities and knowing that it does is seeing it in action. An enthusiastic demonstration, especially if students participate, often can make the difference in winning support. Learning thrives in an exciting environment, and most schools act intuitively upon sensing this.

Becoming skillful simply means learning about telecommunications and using it to serve as a bridge between the why and the how. One "can-do" teacher can convince an administrator who no computer "expert" or salesman would ever be able to sway. One pilot unit that has the students (and their parents) buzzing with excitement can speak louder than all the theoretical good reasons.

STRATEGIES FOR GAINING SUPPORT

Gaining support for telecommunications is accomplished in a number of ways. Which one, or combination, will work depends on each particular school's situation.

Demonstration Approach

The fastest strategy is to demonstrate telecommunications to the principal. If a telecommunications system is available, ask the principal for permission to use a telephone line in order to conduct a demonstration unit. Be prepared to discuss the goals and objectives of the plan, and to explain how it works and what the teacher and the students are doing. See that the students are knowledgeable about the purpose of the activity and are able to explain it to others. When the students have learned the skills and the unit is going smoothly, invite the principal to visit or participate in a class "show and tell." Offer to give a presentation with the principal to the school's parent organization. Showcase the educational advantages and student progress that telecommunications can produce.

Pilot Project Approach

An effective strategy for building a base of support for a new teaching methodology, although somewhat slower than the demonstration, is to carry out a pilot project. A pilot project is a focused, short-term effort to explore the feasibility and effectiveness of a new idea. Such projects permit exploration without the need to commit significant resources until and unless the idea proves to be effective. The focused nature and short time frame of a pilot project do not jeopardize the overall learning environment, yet it still engenders the enthusiasm that successful projects create. In conservative settings, the project can be an elective for students.

Telecommunications lends itself to the pilot project approach. A single system can be set up relatively inexpensively, often without infringing on other school computer use. Group-oriented projects, proven effective in many settings, offer a promising means to showcase the potential for a more extensive program. Special needs or gifted students make excellent pilot groups. They are generally small, already use nontraditional curricula, and tend to have established high levels of student participation. A general expectation exists that successful learning models developed with these students will be picked up by the rest of the school. Successful pilot projects are often merged into the mainstream curriculum in succeeding years with little objection or debate.

Gifts and Grants Approach

School administrators are often very receptive to new methods of teaching if they bring additional resources to the school. Locating external funding sources may be well worth the time and effort. Gifts and grants are available from many sources. Well-known supporters of innovative educational projects include the federal government and its many education-related offices, national educational organizations, state departments of education, and many philanthropic foundations. Other potential sources worth exploring include scientific and technological corporations (especially those with offices in the community), colleges, and local businesses. Fund raisers, school clubs, student activity funds, and departmental budgets are sources closer to home that can often support startup or demonstration telecommunications systems.

Solicitations for most grant funds are highly competitive. A good proposal with clear goals and objectives makes the difference. Such a proposal answers the key questions: what, how, why, who, when, where, and how much. While not always organized in this sequence, grant proposals, however informal, always address these questions. *What* means explaining the overall concept and purpose as well as the educational goals and objectives. *How* suggests including specific plans describing student activities and outcomes. *Why* involves describing the educational problems that the project is intended to solve, or the improvements in learning that are anticipated. *Who* means describing the target group of students and the project leaders' backgrounds so as to demonstrate reasonable qualifications. *When* includes outlining the preparatory, implementation, and evaluative phases of the project. Dissemination plans are expected for some types of grant proposals as well. *Where* involves specification of existing or needed facilities. *How much* requires a detailed budget, including capital expenses, operating or overhead costs, salaries, and miscellaneous expenses.

Most formal grant programs provide application forms that specify how this type of information should be organized. Independent proposals to potential sponsors should address all of the key questions discussed above, following the suggested form and devoting a page or two to each. Cover letters are important in both cases, as are follow-up telephone calls to confirm receipt of the proposal and willingness to respond to additional questions. Once submitted, show an interest in the status of the proposal. It is appropriate to inquire as to the decision timetable and probable date of notification.

The process of developing the proposal should include the school principal and/or curriculum director. It is important that they be co-applicants to ensure that, once funded, the project will be implemented as planned. As a practical matter, it is critical that the roles of each of the participants is clearly delineated and approved. Nothing is sadder than a good idea that cannot be realized because those who should have conducted the project did not obtain the support and approval of the local administrators. Keep in mind that a grant is a means of building local support as much as it is a means of obtaining resources.

Participation Approach

Getting the principal and/or curriculum director online is another approach. As a variation on the "if you can't beat 'em, join 'em" theme, support follows easily if the technology offers something of intrinsic or professional value to those persons whose support is sought. Be thoughtful in determining an application of telecommunications that would be useful to the school's administrators. Explore the possibility of getting a special board set up for the school's administrators with the system operator of a local bulletin board. Train the administrators to use the new telecommunications facility. Electronic messaging has generally taken hold in any organization in which it has been made available. Alternatively, put administrators online with a local FIDOnet "echo" board. School professionals often express a sense of isolation in their jobs; connect them to a wider support group, or log on to SpecialNet or an information utility using ERIC. Show them the wealth of information on educational theory and practice that is just a phone call away, or put them on the local bulletin board where they can participate in the excitement of local computer users. Choose carefully to make this approach useful and collaborative. Use it as a model of how learning can be a lifelong process and how telecommunications helps make that process a reality for more and more people.

Comparison Approach

Developing a data base of how telecommunications is being used by similar schools is another idea. Consult the schools whose projects have been mentioned in this book. Collect news clippings of exciting educational projects. Find out what is being done by nearby school systems. Ask the local Computer Using Educators group to conduct a survey, or check with the state department of education. Every school system tends to maintain a position relative to nearby schools and with schools of similar size and student composition. Some wish to be out in front in most educational developments, some are content to be on par with the others, and some are ultra careful or just cannot afford to make changes. A question that is sure to be asked is, "what are the other schools doing?" Being prepared with the answer demonstrates a sensitivity to this dynamic of school decision making.

FOLLOW-THROUGH PLANNING

Once the support base is attained, a follow-through plan is needed. This plan includes meeting the school's needs, developing a working budget, forming an advisory group of interested people, and publicizing the project. The plan should consist of both curricular and technical elements. The curricular element probably has been implied during the process of building a case for telecommunications. Now it needs to be made specific. The technical requirements need to be specified for both the startup and operational aspects of the plan. A guiding committee, formed to direct the development of telecommunications in the school, gives everyone a stake in its success and serves as a means of determining priorities and resolving any conflicts. Publicizing the school's telecommunications activities keeps enthusiasm high and interest strong.

Curriculum Planning

Curriculum planning for telecommunications takes four aspects into consideration: relevancy, progress, participation, and timing. To become a successful part of the school curriculum, telecommunications needs to be relevant to the ongoing program. Telecommunications should offer a better way of teaching what is already taught or should replace outdated methods. Many curricular ideas have already been outlined in earlier chapters. They provide a wide sampling of applications across the school curriculum that other teachers have found relevant. Telecommunications is integrated much more successfully if it is used as a new teaching method rather than another subject for study. No school curriculum has space for another subject, but every school curriculum can benefit from a more efficient and effective teaching method.

Progress is the growth of the telecommunications curriculum as skill and interest develop. In all learning, interest and enthusiasm tend to wane as the objectives are achieved. Most users acquire the necessary skills fairly quickly, and are then faced with the problem of what to do next. Once telecommunications skills have been acquired, they must then be used to help students progress and grow in the content fields. Effective long-term use of telecommunications requires the development of projects that continue to expand and challenge students' abilities.

Participation refers to the need for a realistic plan for student and teacher use of the telecommunications system. As with any other teaching methodology, successful classroom management is the key to success. Access to the telecommunications system is the key aspect of this new methodology. Does it work best if the teacher operates the system? Should students operate online or prepare files on other computers for transfer? Should students work in groups, pairs, or alone? What curriculum resources are needed to support the online learning activities? Should students have their own telecommunications disks or use a common class system? What are other pupils doing when the rest are online? Should the telecommunications system be one activity center among many, or the main learning station? Do children learn telecommunications skills individually, in groups, or as a class? These aspects of unit planning need to be worked out in order for telecommunications programs to be successful. For most schools, the system represents a "scarce resource." Thought of in this way, decisions can be made more easily about what takes place online, when it happens, and who does it.

Finally, it is important to talk about time. Telecommunications is no different from any other form of teaching. Initially, time is needed to introduce students to the technique. As with anything that is new, it takes longer to teach that first unit the new way. Resisting the temptation to go back to old methods pays dividends before long. Once students have developed skill with the telecommunications system, teachers report that its efficiency more than makes up for the startup time it required.

As a mainstream teaching technique, time has to be allocated, as experience dictates, for online sessions. Not incidental to the telecommunications process, time needs to be built into the unit of study, just the way preparation and cleanup are for many other units. In fact, it is helpful to think of this time as "preparation and cleanup." As with any concrete teaching medium, telecommunications requires a certain amount of time to organize and manage the materials of the medium—computer, software, and files in this case. Finally, follow-up time is an important feature of telecommunications. As an interactive teaching technique, students need to have the time to participate. One of the most exciting educational qualities of telecommunications is the desire students often feel to respond to electronic communications immediately. Allowing time for these "teachable moments" is an important aspect of this teaching methodology.

Technical Planning

When the support has been developed and the curriculum goals identified, there are the technical requirements for making telecommunications work in the school. Three technical aspects need to be considered: telephone access, system setup, and teacher training. Telephone access is essential to a successful telecommunications system; without it, a full-featured telecommunications curriculum cannot exist. The first consideration is the kind of phone line. A direct outside line installed in the classroom is the best alternative, as it supports all forms of telecommunications and does not tie up a school telephone. Calls can be made to both local BBS and information utilities. As telecommunications experience grows, a direct line can support a school BBS, should a desire for one emerge.

Telecommunications can be conducted through an already installed telephone. To work, this phone must be within 100 feet of the telecommunications system, as that is the limit on effectiveness between modem and telephone jack. Again, a direct outside line is the preferred choice. Remember that this telephone cannot be used for regular calls while it is being used for telecommunications. If the phone is often used for other school business, it is necessary to schedule the telecommunications sessions. The system cannot, in this case, support a BBS, since this requires a dedicated outside line. If the school phones all connect through an automatic (dial 9 to access local outside lines) switchboard, long-distance calling usually is not possible, nor can it support a BBS. However, use of local telecommunications resources is still possible.

As indicated, a direct outside line in the classroom is the best alternative, but getting it may require careful negotiation with the principal due to the anxiety that exists about "control." Most people have not yet begun to see the telephone as an "access port." Whether expressed or not, school officials worry about unauthorized use of the phone line for long-distance calls or students' personal calls. Bob Shayler's experience at San Leandro High School, San Leandro, California, is worth sharing on this point:

> We have had some personal calls made by teachers and some outgoing calls made by students (both unauthorized) but no serious problems. The outgoing modem calls were stopped by removing the terminal programs from the system hard disk (requiring students to check out a disk for telecommunications purposes). Most personal calls were stopped by having a student act as BBS monitor each class period. (Shayler, n.d.)

The best approach is to point out that no telephone set is needed and that modems do not support voice calls. If necessary, locate the telephone where the students are not, such as in the teacher's room or school office. It might also help to consult with the local telephone company for suggestions on how to secure the line and the availability of cutoff switches. In the end, only responsible use decreases anxiety. Requesting a trial use period through a school phone to demonstrate that the system is controllable is one means of working up to a direct outside line in the classroom. If support for the importance of educational telecommunications is developed, it is usually possible to find a way to work through these concerns.

After telephone access is determined, computer requirements, modem type, and telecommunications software needs come next. With respect to the computer, it must be able to interface with the modem of choice. Determine the brand, memory size, and interface options of the computer. Consult with the local computer representative about the compatibility of this computer with various modems. If a computer does not support a modem connection, it is worthless for telecommunications. Once you have selected a computer that supports a modem, obtain any interface cards and cables that go along with the modem. Unfortunately, these generally must be purchased separately.

The two primary considerations for the modem are whether it is internal or external and the baud rates it supports. Internal or external is a matter of what is best at the location and with the particular computer. Internal modems save space, have fewer cables, and do not require an additional electrical outlet. External modems are not computer specific, but require an additional interface card. Computers without internal interface slots cannot use internal modems. The use of 300, 1200, or 2400 baud modems depends on intended use. Most BBS systems support 300 baud. It is also the most reliable rate for ASCII character transfer and is not as susceptible to noise interference. Baud rates of 1200 and 2400 cut down on the time required for file transfers and connect time to information utilities, although charges are sometimes higher when data is sent at high baud rates. Having a modem that supports 300, 1200 and 2400 baud gives the option of using slow or fast exchange, but these modems cost more.

Other features require knowing the technical requirements or intended operation. For example, to set up a BBS, a modem that can answer the telephone automatically is needed. If compatibility with many communications programs is desired, consider a modem that responds to the Hayes Smartmodem command set. For new users, a 300/1200 baud, Hayes compatible modem that can be interfaced to the computer is a good choice.

Third, choose a communications software program that is compatible with the hardware. As discussed in Chapter 2, ease of use is a key feature and hardware compatibility is an absolute requirement. Commercial products generally list their compatibility features; public domain and "shareware" products do not. When a person is not technically inclined, or does not have a friend who is, or does not have the time to make "free" software work, a commercial product pays for itself quickly. As Martha Stanton, Curriculum Resource Coordinator for the Lexington Public Schools, Lexington, Massachusetts, can attest:

> School libraries access the public library's online catalog, using the Institutional Loop of the Town's cable television system for data transmission, thereby eliminating on-going communications costs. Access is by author, title, and subject as well as language of publication and topics in the community resource file. Communications software was developed by Lexington High School students. A problem we've encountered as a result has been getting enough student time assigned to debugging the program. (Stanton, n.d.)

Finally, for widespread teacher participation, a teacher training program is absolutely critical. Telecommunications is a new and intimidating concept for many people. Many teachers shy away from new technology. If telecommunications is to become a part of the school curriculum, it is necessary to teach the teachers first. One-shot afternoon workshops do not do it. They are good introductions, especially if they are a "show and tell," but teachers need hands-on experience in a "safe" setting. A "safe" setting is usually among a gathering of other novices and uncertain teachers. Hands-on training means letting them go online themselves with assistance right at hand. Working with small groups and rotating everyone through an online session is an excellent approach.

The training should follow a progressive plan. In fact, this training can prepare teachers to do the very same things with their students. Depending on the timetable for introduction and the availability of the resource to the various classrooms, a long-term schedule of training and help sessions can be planned that guide teachers through the process of integrating telecommunications into the curriculum. The goal should be the development of confidence and skill with operating the system, leading to discussions and curricular planning for use of the tool in the classroom. In this fashion, teachers can be expected to become as excited as the students once their special training and support needs are met. This support must be included in the planning for telecommunications as explicitly as the curricular and technical aspects are.

The Working Budget

The goal in budgeting is matching costs with curriculum objectives. This is a dynamic process with adjustments on both sides as experience is gained and goals evolve. Working within the budget builds good relationships with the school's decision makers and shapes a climate that reflects the importance of telecommunications as it becomes a part of the curriculum. Therefore, efforts to contain costs both increase the effectiveness of the budget and are seen as responsible action, lending more weight to new estimates and statements of need in future budget cycles.

Important factors that need to be considered in a basic working budget include:

Startup Costs:
Computer
Monitor
Memory
Disk drive
Modem
Interface card (if needed)
Cables
Communications software
Telephone line installation
Information utility one-time membership fees

Ongoing Operating Costs:
System repair and maintenance
Software replacement and upgrades
Teacher training expenses
- instructors
- materials

- utilities
- janitorial

Monthly telephone line fee
Private line fee (if necessary)
Estimated connection costs (estimated hours of use times rate per hour)
Monthly costs for information service

- long-distance fee
- access fee
- connect time fee
- information fee
- minimum fee

Disposable password accounts

These costs can be contained in a number of ways. The basic goal is to improve the efficiency of online time. One method is the use of offline telecommunications simulators to teach online skills. Until people have demonstrated facility with the hardware/software system, they should practice offline. Simulation programs to accomplish this are discussed in Chapter 3. Another approach is to connect two computers together through modems. Modems can be connected to each other with telephone extension cords. Students can be in the same class, or in two different rooms, and can use the real communications software to "chat" with each other as if they were online. If one computer is running a BBS, there is no difference between this and calling the system through the phone line, except that dialing a number is not necessary.

Another method of cost containment is the use of downloading, discussed in Chapters 2 and 8. The online session lasts only as long as it takes to transfer the necessary files to disk. Using 1200 or 2400 baud cuts connect time significantly. Students then retrieve the downloaded files after they are offline. Furthermore, this retrieval process can use computers other than the one used for telecommunications, freeing it for the next student. Finally, as more and more educators use information services, we can expect to see the development of billing options that reflect the needs of schools to control costs. One example that is available now on the Einstein system is the "disposable password" (see Chapter 5 and the Resource Section). With Einstein, schools can purchase blocks of passwords in advance. These are given to students when they are ready to go online and are good for one session on the system. The concept offers schools an effective option, and similar alternatives will undoubtedly be developed by other information utilities.

With careful planning, telecommunications can become an effective teaching technique in any school. Inherently motivating and definitely state-of-the-art, telecommunications puts students in touch with their futures. As a medium of communication, it supports the fundamental goal of helping students become literate and fluent. As a method of expression, it offers another arena where students can develop their skills at thinking and expressing themselves.

REFERENCES

Shayler, B. San Leandro High School, San Leandro, CA.

Stanton, M. Curriculum Resource Center, Lexington Public Schools, 17 Stedman Road, Lexington, MA 02173.

PART III

ADVANCED TOPICS

8

Technical Considerations

Many newcomers to telecommunications find the concepts and terminology mystifying at best, and intimidating at worst. A detailed knowledge of this information is not necessary before being able to get online successfully, but more understanding is useful as experience and skill increase. A basic knowledge of certain key terms can be helpful when making curricular plans and purchasing equipment. Most of the terminology is a shorthand way to describe the concepts and capabilities of telecommunications systems. This chapter describes many of the major aspects of telecommunications through its vocabulary. Chapter 9 discusses the use of more advanced technologies in education, presenting yet another level of concepts, vocabulary, and pedagogic potential.

In the belief that technical terms and concepts make more sense when explained in the context of their contributions to functional systems, the information in this chapter is arranged along operational lines. The telecommunications process, and its related language, is described from inside the computer working out. The chapter begins with a consideration of the hardware components of a telecommunications system, describing in turn computer, interface, and modem technology essential to the telecommunications process. In essence, we follow a byte of computer information through the computer hardware and peripheral devices that prepare it for exchange with another computer.

The next topic looks at how communications software eases the job of setting up the telecommunications hardware. The use of the control mode for terminal emulation and protocol settings is taken up first, followed by a consideration of less critical settings and control commands. Completing this section is a description of how data is actually exchanged.

The chapter concludes with a discussion of communications software features considered at three levels: essential, convenient, and advanced. The essential features are those aspects of software that make for friendly telecommunications. These constitute what the authors feel are the basic set of functions that a telecommunications program should perform. The convenient features are those functions that greatly ease the user's job in managing the flow and use of exchanging data. Persons who use telecommunications on a more than occasional basis generally want their communications software to have these functions. The advanced features are those that are important primarily to serious personal users and those contemplating business use.

HARDWARE-RELATED INFORMATION

About Binary

Let's begin by considering the underlying technical nature of the computers that are communicating with one another during the telecommunications process. Computers are binary machines, which means that they represent information in terms of ones and zeros. By putting ones and zeros together in larger sets, the resulting binary numbers can be made to represent virtually anything that a person might wish to know: addresses, grocery lists, term papers, computer programs, mathematical calculations. Most personal computers use binary numbers consisting of 8 or 16 ones and zeros, called bytes. The individual ones and zeros are known as bits. Because a computer is an electrical device, these 1 and 0 bits exist as voltages inside the machine; specifically, a 1 is represented by a voltage that is greater than 3 volts, while a 0 is represented by a voltage of less than 1 volt.

Interface Ports

In order for one computer to communicate with another, there must be a way to send and receive these electrical bytes of information to and from the outside world. Most computers provide this access through what is called an interface port. An interface port is created by adding another circuit—an interface "card"—into one of the computer's interface slots. The user can then attach peripheral devices (like modems and printers) to the interface port with a special cable that establishes the electrical connection for the exchange of information bits.

RS-232C Interface

The most common interface port for telecommunications is called RS-232C, for Recommended Standard #232 Version C, published by the Electrical Institute of America (EIA). It refers to a standard set of voltage levels and associated logic conventions that are compatible with any peripheral device designed to work with an RS-232C port. This interface is necessary because the computer voltage levels are too small to work at any distance outside the computer itself. The RS-232C interface port changes the computer's 3-volt 1 bit to more than 15 volts and the less than 1-volt 0 bit to less than −15 volts. These greater voltage levels are able to carry the binary information to peripheral devices such as modems and printers over cables up to 6 feet long.

Parallel and Serial Data

Inside the computer, all the bits in a byte of binary information are handled simultaneously, like marchers in a parade walking abreast. This is known as parallel data. Very few peripheral devices accept parallel connections (Centronics and Epson printers being major exceptions). To do so requires a cable with a separate wire for each bit of data. This type of cable is called a ribbon cable, and for bytes containing more than 8 bits they become increasingly difficult to handle. Also, the cable must be very short if the small computer voltages are to reach the device reliably.

Telephone lines are serial in nature, that is, they can only handle one bit of information at a time. A second function of the interface port, then, is to change the data from parallel to serial when sending, and from serial back to parallel when receiving. The interface port acts like a turnstile, funneling the outgoing parallel marching bits into a single file line so that they can fit through the serial telephone line, and reassembling the incoming line of serial bits into parallel formation for the computer.

Modem

For telecommunications, the serial bits formed by the interface port are sent along a special cable to a modem. A modem is a peripheral device that can be attached to an RS-232C interface port. It translates the voltages representing bits of computer information to and from audible sounds for transmission over ordinary telephone lines. The term is derived from the two functions it performs: MOdulate, meaning to add information to a communications channel, and DEModulate, meaning to extract information.

Acoustic and In-line Modems

Modems can be connected to telephone lines in one of two ways. The older method, now used mainly for portable equipment, has two rubber cups containing speakers that cradle a telephone handset. This is known as acoustic coupling and works by having one of the modem's speakers "talk" by making sounds for the telephone mouthpiece while the other listens to "hear" sounds from the telephone earpiece. Most modems in use today, however, connect directly to the telephone lines. Known as direct or in-line modems, the audible tones remain entirely in their telephone analog voltage form.

Internal and External Modems

Based on how they are attached to a computer, two types of direct connect (in-line) modems have been developed: internal and external. Internal modems generally look like printed circuit cards and are designed to plug directly into the interface slots of a computer. These modems combine the RS-232C interface on the same circuit board with the modem. The telephone connections are made through the back of the computer directly to the modem circuit card itself. The advantages of this form of modem are its one-piece construction and use of the computer's power supply.

External modems operate outside the computer itself and require a separate interface card or dedicated (built-in) computer modem port. The telephone connections are made to the modem, and the modem is then connected by a special cable to the computer port. Generally, a separate electrical outlet is needed for power. The advantages of this form of modem are its compatibility with many types of computer interface and port designs and the option of omitting the RS-232C interface if the computer already provides a modem port.

Command Set

Once the modem is properly connected to the computer through an interface port, it is possible to use it to establish a telecommunications connection with the host computer. This is usually accomplished by typing instructions at the computer keyboard. By typing special character sequences or "strings," using a command set that the modem understands and interprets as directions, the user can control the modem's operations.

The command set provides both direct manual keyboard control of the modem as well as a means of software control for sophisticated communications programs. The "AT" or Hayes compatible command set was developed by the Hayes Corporation to control the operation of their Smartmodem and has become a standard for most of the modem industry.

For example, a Hayes compatible modem can be commanded to place a telephone call by typing AT DT (or DP) followed by the telephone number and a carriage return. The AT tells the modem that the characters and numbers that follow are a modem command rather than data to be transmitted. The DT tells the modem to dial a touchtone call (DP means dial a pulse—rotary dial—call). The numbers are the tones or pulses that are dialed. The carriage return tells the modem that the command string is complete. It is a good idea to choose a modem and communications software program that are Hayes compatible to ensure compatibility with many other hardware and software telecommunications components.

Placing a telephone call using a modem's built-in dialing capabilities is one of the most common uses of the modem command set. When the dial command is issued, the modem begins to place the call and establish telecommunications with a host computer. The first action is to turn on the monitor speaker, if there is one, so that the user can listen in on the progress of the call. Then the telephone line is activated and the telephone system dial tone can be heard through the monitor speaker. After a short pause, the modem's autodial circuits place the call using either touchtones or rotary pulses (that sound like clicks). The telephone system's switching circuits are then heard, just as they are when a manual call is placed, setting up the connection to the desired telephone number. When the connection is made, either the ringing tone or busy signal is heard.

If the ringing tone is heard, the host computer generally answers the telephone within six rings. When it does, a high-pitched tone is heard on the line. When the modem detects this high-pitched tone, it responds with a tone of its own, and turns off the monitor speaker. At this point telecommunications has been established, and all further information about the progress of the call is seen on the computer screen.

If a busy signal is heard, the user has to cancel the call and place it again. Hanging up the telephone line is accomplished by typing AT H followed by the return key, the hang-up command for Hayes compatible modems. The monitor speaker is then turned off and the telephone line is deactivated or hung up.

Carrier Signal

If the host system answers the call, the steady high-pitched tone that it sends out is known as the carrier signal. Telecommunications contact is established when the host computer's modem "answers" the telephone call, begins sending out the carrier tone, and is responded to by the calling modem. In fact, calling modems are designed to detect and maintain this tone throughout the online session to prevent the modems from disconnecting. The carrier signal also acts as a failsafe mechanism to automatically turn the user's modem off and hang up the telephone line if something unexpectedly breaks the connection.

Tone Pairs

The carrier signal sound is not the sound of information being exchanged between the two computers. To exchange information, modems generate special sounds called standard tone pairs. These special tones are known as the Bell 103 and Bell 212A standards for data transmission over regular telephone lines. They are important because two computers cannot communicate with each other unless they are both using the same set of standard tone pairs. Modem users do not generally need to concern themselves with the compatibility of the tone pair sets. These are associated with the speed of data transmission and are set automatically by most modems when the baud rate is selected. The Bell 103 standard is associated with 300 baud data exchange and Bell 212A with 1200 baud. Baud rate is discussed in some detail later in this chapter. Simply put, the baud rate is the speed of data transfer expressed in terms of the number of bits per second.

Answer and Originate Modes

Because a telephone can both send and receive information at the same time, it is important for accurate data transmission that the two modems be set to send and receive opposite tone pairs. When making the telephone connection, the calling modem needs to be in the originate mode while the answering modem needs to be in the answer mode. That is to say, the originating modem needs to be set to send tone pair A and listen for tone pair B, while the answering modem needs to be set to send tone pair B and listen for tone pair A.

With most modems on the market today, the mode is automatically set to the correct originate mode when a call is dialed. Some of the older acoustic modems, such as the Novation CAT, have a switch that the user must set to originate mode before the call is manually placed. Only when two people with personal computers decide to exchange files through direct computer-to-computer transfer do they have to concern themselves with setting modems to the correct mode, and then only the one that is the answering modem. For more information, see the description of "chat mode" later in this chapter, or consult the modem's user manual.

Handshaking

As previously described, the modem operates by converting serial bytes of binary information into a sequence of sounds that represent 1 and 0 bits. A modem must not accept a new character from the computer for data transmission until the last character has been completely sent. It must not transfer an incoming character to the computer during data reception until all 7 or 8 bits of the character have been received. The process that controls this flow of data between modem and computer is known as "handshaking."

Handshaking is conducted by way of the cable that connects the computer interface and modem. This cable contains some special wires that are controlled by the RS-232C interface and that tell both computer and modem when the other is ready. In this way, incoming and outgoing data can be smoothly transferred.

The RS-232C standard defines ten handshaking signals of which six are essential to smooth telecommunications:

1. The *Transmit Data* (TD) signal tells the modem that character data is ready to be sent to it from the computer over this line.
2. The *Receive Data* (RD) signal tells the computer that the modem has character data ready for it on this line.
3. The *Data Set Ready* (DSR) signal tells the computer that the modem is ready and waiting for command set instructions and that it is not in test, dial, or talk mode.
4. The *Data Terminal Ready* (DTR) signal tells the modem that the computer is ready for telecommunications and that it should switch to the telephone communication channel.
5. The *Carrier Detect* (CD) signal tells the computer that the modem is receiving a carrier signal from a remote modem on the telephone communication channel.
6. The *Signal Ground* (SGND) signal provides a reference point against which the polarity (plus or minus) and magnitude (amount) of the signal voltages representing character data can be measured.

The other four handshaking lines available with the RS-232C standard are sometimes used for more sophisticated communications control:

7. The *Request to Send* (RTS) signal tells the modem to switch to transmit mode. When using half duplex, this signal inhibits the receive mode.
8. The *Clear to Send* (CTS) signal tells the computer that the modem is ready to accept data on the Transmit Data line.
9. The *Ring Indicator* (RI) signal tells the computer that a telephone ringing signal is being received on the telephone communications channel.
10. The *Protective Ground* (PGND) line provides an electrical ground for the modem.

To summarize, then, a telecommunications system necessitates a special set of hardware devices. These permit a computer to convert bytes of information into electrical signals that can be sent over cables and telephone lines to other computers. Chief among these hardware devices are an interface port and a modem. The interface port serves to convert parallel bytes of computer information into serial streams of individual data bits (and vice versa), and change the voltage levels of these bits between the higher levels required by peripheral devices like modems and the lower levels used by the computer. The modem:

- manages communication between the user's computer and itself,
- establishes contact with the modem of a host computer system on command,
- modulates the telephone line with tones that represent the serial bit stream of information to be transmitted, and
- demodulates received tones into a serial bit stream of information for the user's computer.

When properly set up and functioning, this hardware establishes a very basic telecommunications system. Controlled by the modem's command set, it is possible to use such a basic system for a great many telecommunications purposes. Manual control, however, requires a great deal of effort on the part of the user. In order to ease the control burden, and thereby facilitate the telecommunications process, computer programmers have developed special programs that can use the modem's command set to control its operation automatically. The extent to which these special programs assist users in achieving their telecommunications needs determines how "user friendly" the telecommunications process is. Let's now turn our attention to the software aspect of telecommunications.

SOFTWARE-RELATED INFORMATION

Communications Software

Communications software, sometimes known as a terminal emulation program, is the computer program that makes it possible for a user to have the computer do much of the work of controlling the modem. When using the computer for telecommunications, the communications software has to attend to at least two sources of incoming information, the keyboard and the modem, and to determine which of a variety of different locations, including screen, modem, printer, and disk drives, to send outgoing information. Communications software, like *Apple Access II* and *Procom Plus*, makes it easy for both the user and the computer to know how to set up the modem, when to send and receive data through the modem, what to do with information typed at the computer keyboard, where to store and retrieve data files, and how to display information. The communications software takes charge when it is loaded (booted) into the computer and

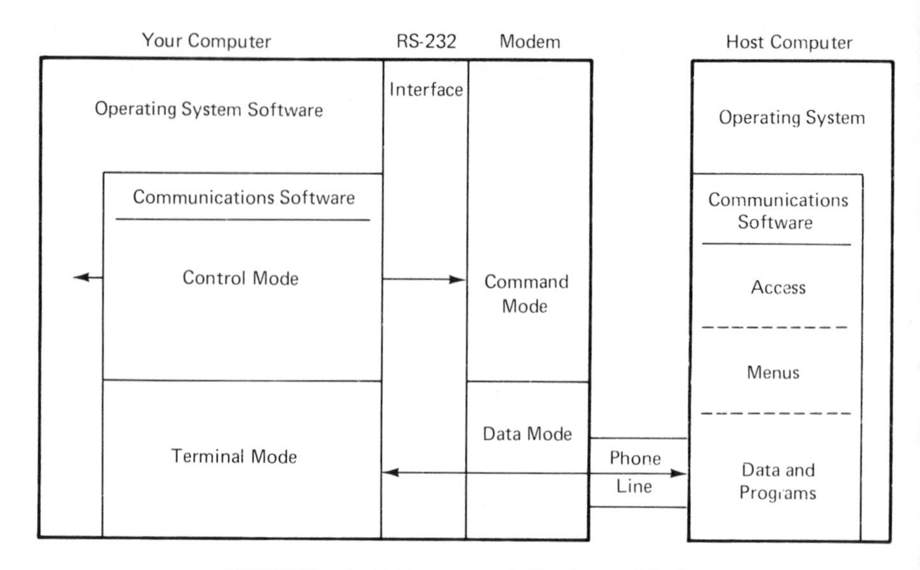

FIGURE 8.1. Software and Hardware Modes

run. Generally, when this occurs the computer is placed in what is called the control mode in preparation for telecommunications.

Control Mode

Communications programs generally possess two operating modes, control and terminal. In control mode the user is interacting with the communication software rather than the modem. Control mode permits the user to be in control of their telecommunications system. Most communications programs provide menu options and various prompts and responses in control mode that allow the user to set the configuration of the modem so that the system is compatible with a host computer. The software at this stage sets the type of terminal emulation and the modem's operating parameters, called protocols, and is invisible to the user. The modem is in its command mode and is not yet connected to the telephone line. Generally, control mode also provides utility routines that permit the user to manage any printers, disk drives, and file storage and retrieval devices that may be a part of the telecommunications setup.

Terminal Emulation

Preparation for telecommunications requires two main software activities, describing the data display format and setting up the modem parameters. The former is known as setting the terminal emulation, while the latter concerns the modem protocols. Let's consider terminal emulation first.

Largely because today's telecommunications systems have evolved from the mechanical teletype networks of years gone by, most terminal programs operate as though the computer were a *teletype machine*. Known as TTY terminal emulation operation, the terminal program assumes that the incoming data is to be displayed in sequential order on the screen. Cursor control is limited to backspacing, and only line feeds that advance the cursor's position downward on the screen are recognized. In essence, a TTY (teletype) is a line-oriented device.

Also, mechanical teletype machines did not have any built-in memory or control capabilities. They were controlled entirely by the incoming signals, and are described in computer terms as "dumb" terminals. When a terminal program emulates the teletype (TTY) mode, it causes the user's computer to behave as a "dumb" terminal. None of the computer's built-in memory or control features are utilized.

Computers have a different developmental history, and many offer complete control over cursor position on the monitor screen as well as the ability to permit host systems to use their built-in memory and control features. That is to say, many computer terminals have the capability to act as "smart" terminals, or screen-oriented devices. Some host computer systems operate in screen-oriented mode. While these are not yet common in public telecommunications systems, two widespread standards that are beginning to appear are the ANSI (American National Standards Institute) VT100 and VT52 standards. The VT100 and VT52 are computer terminals developed by the Digital Equipment Corporation. VT stands for video terminal and means that the computer is indistinguishable from one of the host system's own directly connected terminals. Many sophisticated communications software programs can now emulate either or both VT100 or VT52 operation. When using these telecommunications systems, certain special key combinations known as "escape sequences" can cause the cursor to move in other than a line-oriented fashion. This is an advanced telecommunications topic that is best learned by consulting the host system's operator for information and instruction.

Protocol

Protocols are the other primary responsibility of communications software in control mode. Protocols refer to the settings of the data exchange conditions that the modem uses. They must be set to match those being used by the host computer or else telecommunications is difficult, if not impossible. One convenient feature of communications software is its ability to permit these protocol settings to be made initially and then saved for automatic use in future sessions. While it is certainly possible to set these protocols manually with the modem's command set (Hayes command set), the ability to set, store, and edit the protocols is a primary measure of the friendliness of a terminal emulation program. The single greatest advantage of a communications program is its ability to use the power and features of the computer's memory and data storage to eliminate the need to repeat modem protocol commands each time the system is used for telecommunications.

Three protocol settings are critical and should be capable of being set and edited by the communications software: transmission speed, data format, and data flow. Transmission speed is known as baud rate; data format is comprised of number of bits per character, parity, and number of stop bits; and data flow is either one-way half duplex or two-way full duplex. Each of these parameters is discussed in detail in the paragraphs that follow, but for operational purposes the most common settings are 300 baud, 8 bits per character, no parity, 1 stop bit, and full duplex (known as 300, 8N1). These settings have become nearly standard for telecommunications with personal computers. If they do not work, 300 baud, 7 bits per character, even parity, 1 stop bit, and full duplex (300, 7E1) is also in widespread use.

Baud Rate

The baud rate protocol is a measure of data transmission speed. It is the rate of bits per second being sent over the telephone line. For example, 300 baud means that 300 bits can be sent every second by the modem. The term is derived from the name of Emile Baudot, who invented the 5-bit teleprinter code in 1874 for use by mechanical teletype machines. Expressed in terms of meaningful information, the number of baud divided by 10 is roughly the number of characters that can be sent per second. This is because personal computers generally follow the ASCII (American Standard Code for Information Interchange) convention, using 8 bits to represent characters. When sent over the phone line an additional bit called the "start" bit is added to the beginning of each character to signal the host computer's modem to start collecting the following 8 bits as ASCII data. A "stop" bit is added to tell the host computer's modem that the character is complete. Thus, each character generally consists of 10 bits, making 300 baud roughly 30 characters per second. This varies slightly with other protocol settings, such as 7 bit characters or the use of a parity bit or 2 stop bits. It is also different than the "true baud" measure of meaningful information transfer that is affected by the error rate and the need to send data more than once in order to ensure successful data exchange.

When baud is considered in terms of what is seen on the screen, about 2.5 seconds is required for a full line of 80-column text at 300 baud, with a full 24-line screen taking up to 1 minute to be transmitted. This is not particularly fast, especially if long distance telephone charges are being incurred during the data transmission. Efforts have therefore been made to use higher speeds, and 1200 baud (120 characters per second; full screen in less than 20 seconds) has become a popular option on many computer bulletin boards and information utilities. Recently, 2400 baud (240 characters per second; full screen in less than 10 seconds) capabilities have been added to many systems.

For telecommunications using local computer bulletin boards and the sending of messages that are typed at the keyboard as they are being sent, 300 baud is plenty of speed and 1200 baud is a pleasant convenience. The main advantage of higher baud is the increased speed of data exchange. This becomes important if the transfer of prepared files is frequently done. However, there is a trade-off between savings in connect time charges and the higher access charges levied by information utilities for using higher baud rates.

Bits per Character

The modem also needs to know how many bits comprise one byte of computer information. It needs to know this to reassemble the stream of individual bits into a functional byte. Character data—numbers, letters, and punctuation—are represented in personal computers in a standard form known as ASCII. The ASCII character standard defines the 8 bits that comprise a byte of computer data in 128 unique patterns such that every key on the keyboard has a distinct binary number (0 to 127). In doing this, only 7 of the 8 bits in a byte are used. The most significant bit is undefined at present.

Other computer data representations such as numerical quantities, program operation codes, memory addresses, and graphic symbols use all 8 bits. No standard meanings exist for these bytes; each is defined in the context of the program within which it resides. As a result, all 8 bits in each byte convey essential information. Furthermore, while this type of data can be sent and received, the user's personal computer may not be able to use it. This is not text data, although the communications software interprets it as such. The characters displayed on the screen during the data transmission look like gibberish. The data is useful only if the same type of computer, operating system, and program that created the data is receiving the data.

When exchanging data by way of telecommunications, it is important to have the correct *bits per character* setting. A setting of anything less than 8 bits per character works only if the host computer is using the same setting and the data is accurately represented with fewer bits. A setting of 8 bits per character is absolutely essential when binary file transfers of data other than ASCII text are being made.

Start Bit

Modems universally add a bit known as a start bit to the beginning of each character to signal the receiving modem to switch to its data reception mode. Since this signal is now universal, it is no longer necessary to specify this setting for the modem, and most communications programs do not request a parameter.

Stop Bit

As a signal to the receiving modem that a character has been completely sent, one or more bits called stop bits are added to the end of each character. When the stop bit is detected by the receiving modem, its operation is placed into a loop that looks for the next character's start bit. The use of a single stop bit has become nearly standard in personal telecommunications, but $1\frac{1}{2}$ or 2 stop bits are not uncommon.

Parity

The *parity* protocol setting specifies a simple form of error checking. It is intended to allow communications software to determine whether a character has been received correctly. Most personal telecommunications systems have dropped the use of parity error checking, so a setting of *none* is most common today. Those systems that still use it generally use the *even* parity setting. Parity works by setting the binary state of an additional bit according to one of five rules. Odd parity sets the parity bit whenever necessary so that the number of binary 1s in the transmitted character (including the parity bit) is always odd. Even parity sets the parity bit whenever necessary so that the number of binary 1s in the transmitted character is always even. Mark parity always sets the parity bit to the binary 1 "marking" condition, and Space parity always sets the parity bit to the binary 0 "spacing" condition. "None" parity disregards the parity bit entirely.

Parity checking is a fairly accurate means of making sure that characters are received correctly, but unless both computers are using the same parity rule the data will not be received correctly. Generally, to make use of parity checking's automatic error correction capabilities the communications software must examine the parity information and request the host computer to retransmit any character whose parity bit does not match the rule in use.

Because parity checking uses the eighth bit of each character that is sent, it can therefore only be used when sending ASCII text data. This limitation has led to a decline in the number of systems using this form of error checking, and it has been largely replaced by "Xmodem" error checking that works with both ASCII text and 8-bit binary data.

Xmodem Error Checking

Xmodem protocol is an automatic error checking and correcting method that imposes no restrictions on the type of data exchange and that virtually ensures perfect information exchange. Developed by Ward Christiansen for data exchange between CP/M type disk operating system computers, the method has been adapted for most personal computer operating systems. With Xmodem error checking, the data stream is sent in blocks of 128 eight-bit bytes. To each block is added a three byte "header" at the beginning of the data stream and a one byte "checksum" at the end. Essentially, if no error is found in either the header or checksum, it is assumed that the information block sent in between is also correct. In more technical terms, the header consists of:

1. hexadecimal value $01 (binary 00000001)
2. block number in binary form (i.e., 00010011 = block 19)
3. block number subtracted from 255 (i.e., 11111111 − 00010011 = 11101100, or 236 for block 19).

This permits the first error check to be made on the block number itself. When the three bytes of the header are added together, the answer should be zero (the computer's microprocessor accumulator register should contain zero with the carry flag set). This is because an 8-bit byte can represent 256 unique numbers—0 to 255—with values of 256 and greater, causing the byte to begin counting again from zero and a carry bit to be set. If the sum of the header bytes is not zero, then the communications program requests that the other computer send the information block again.

If the header value is correct, then the checksum byte is used to check on the accuracy of the 128 bytes of data sent in the block. The binary value of each of the 128 data bytes is added to the others by both computers as each is sent. The sending computer then sends its answer as the checksum byte, and the receiving computer compares it to the value it calculated. If they are the same, the data block was received without error; if not, then the receiving computer requests that the data block be sent again.

In this fashion, it is possible to send large amounts of text and binary data between computers virtually error free. For commercial purposes, this is essential if data must be absolutely correct. Without this reliability, telecommunications would not provide an effective communications alternative for government and industry. For personal users, the primary advantage lies in the ability to exchange computer programs—writing and debugging programs is hard enough without having additional problems introduced by errors in the telecommunications process. Besides, the programs are sent in their original binary form, requiring no complex ASCII-to-binary conversions to run them once received (assuming computer and operating system compatibility).

Other Protocols

The preceding protocols constitute the critical settings that must match for both modems when attempting telecommunications. There are, however, other protocol settings that are less critical but may affect the appearance of the information on the screen or the ability of the printer or disk drives to work compatibly during an online session. These settings involve *line feeds* after carriage returns, *pauses* after carriage returns (usually called *nulls*), *Xon/Xoff* (explained later under Duplex), *normal* or *inverse video, wraparound, tab stops,* and *duplex*. If the modem or communications software does not permit control of these settings, and it is important that they be adjusted, many host computer systems have a menu choice called something like "status," "user," "account," "examine," or "width" that permits control of how the host computer sends information.

Line feeds after carriage returns determine how text data is displayed on the screen. If the host system does not automatically issue a line feed after each carriage return, all the received data is displayed on the same line since each line of received data overwrites the preceding line. If this happens, the host system may have a menu option that instructs it to begin sending line feeds. If not, check the terminal emulation software or modem user's manual to determine if the terminal can add this command. On the other hand, if all text is appearing double spaced, both the terminal program and the host system are issuing line feeds after each carriage return. To obtain single-spaced text, turn off the line feed option on either system.

As a way to maintain compatibility with mechanical display systems like printers, many host systems can be instructed to insert *nulls* into the text stream. Nulls are pauses or waiting periods that the host system uses after each carriage return to allow the mechanical system time to get ready for the next line of data. Generally this pause allows time for the print head or carriage of a printer to move back to its starting position on the left side of the carriage. If not all the data appearing on the screen is being printed, some nulls need to be inserted by the host system. Another indication that nulls are needed is the regular loss of one or more characters at the start of every printed line.

Xon/Xoff is a standard way to start and stop data transmission. This permits either automatic (software) or manual control of the online data stream. Special control keys are used as control signals, usually Control-S for Xon and Control-Q for Xoff. Full duplex operation is required to use Xon/Xoff. The use of these control signals is explained later under Duplex.

Normal or *inverse video* options let the user determine whether text and data are displayed as dark characters on a light background (normal), or as light characters on a dark background (inverse). This setting is a personal preference and does not affect the data.

The *wraparound* option determines whether incoming data that exceeds the screen width is displayed or not. With no wraparound, when the cursor reaches the right side of the screen it remains there until the next carriage return. As a result, all characters received after the cursor reaches the margin, but before the next carriage return, are not displayed. With wraparound, when the cursor reaches the right margin, a carriage return and line feed are automatically issued and the incoming data is continued on the next line. Many communications programs today have wraparound activated as a fixed feature.

The *tab* character can be transmitted like any other ASCII character. This permits the transmission of data in columns. The interpretation of the tab character, however, is largely the responsibility of the user's terminal emulation program. When expecting data containing tab characters, be sure that the communications program permits locating the tab stops appropriately.

Duplex

Duplex refers to the flow of information between the two computers. Since telephone lines can both send and receive information at the same time, it is possible for both computers to do so. Simultaneous data transfer of this type is known as "true full duplex." To use true full duplex, both computers must be using very sophisticated communications software programs. Most BBS systems do not offer this capability.

BBS systems do, however, generally offer "full duplex," which most communications software programs can use. In full duplex mode, while one computer is sending data, it can also receive special control signals known as Xon/Xoff from the other. This permits the receiving computer to control the flow of information and permits it to process the received data without losing any incoming information. This most frequently occurs when the received data is being automatically stored to disk and the computer must stop receiving data to manage the file storage process, or when the printer is running simultaneously and cannot keep up with the incoming data. When a user presses (or the communications software sends) the Control-Q key combination this stops—Quits—the data from being sent. A Control-S signal Starts it again.

The other use of full duplex is to "echo" the received characters back to the sending computer. These echoed characters are the ones displayed on the sending computer's screen, giving the user an informal sense of the accuracy of the transmission. This error information is only informal because the character has been sent across the telephone lines twice and you cannot be sure where an error occurs. When using full duplex, turn off the "echo character" option of the communications software so that double characters will not appear on the screen as both computers echo characters.

Some older BBS systems cannot receive Xon/Xoff signals while they are sending data. These systems must operate in half duplex, wherein the data flow can be only one way at a time and the receiving computer has no way to communicate with the sender until the transmission is complete. This is similar to a home intercom or to citizens' band radio where only one person can talk at a time. In half duplex, the sending computer cannot receive characters echoed back to it, so the user must turn on the echo character option of his or her telecommunications software to see what the computer is sending.

Terminal Mode

As soon as the communications software has configured the modem and computer for telecommunications, it is ready to make contact with a host BBS and exchange data. If the communications software makes use of the computer's memory and storage devices to save the configuration settings, future online sessions can begin at this point. Since the communication software's control mode does not make use of the modem for data exchange, a command must be issued to place the program into terminal mode.

Control mode always provides an option that causes the communications software to enter the terminal mode (in fact, some programs such as Procom Plus boot up directly into terminal mode). In terminal mode, everything the user types at the keyboard is sent directly to the modem. The modem command set (Hayes command set, if the modem is Hayes compatible) is used to instruct the modem to switch between its command and data modes. The two basic commands are "dial" and "hang up." Dial causes the modem to call a telephone number and switch to data mode if the call is answered. Hang up causes the modem to "hang up" the telephone line, ending the online session, and switch back to command mode. In many communications programs, entry to terminal mode is made by way of the "autodial" option, wherein the program automatically reverts to control mode if the call is not successfully made.

When in terminal mode, certain of the communication program's control mode commands can still be used with special key combinations to give the user control of the computer and its peripherals while online. It is also possible to leave the terminal mode and return to control mode to format a save disk or change printer settings, for example. Exercise caution when doing this, however, because leaving terminal mode usually does not automatically end the telephone call. To end an online telecommunications session, the user must log off the host system or issue a command to the modem to hang up the telephone. Check the communications software's user manual for details.

Command Mode

When a modem is first turned on, it is activated in what is called the command mode. In this mode it is waiting for the computer to send it command strings that will instruct it to perform its functions. All Hayes compatible modems wait for command strings that begin with the letters AT. Other command sets are used by some modem manufacturers, but even these will have some special combination of letters, numbers, or symbols which will indicate to the modem that a command is being sent to it.

As previously discussed, the two basic modem commands are the "dial" and "hang up" instructions. However, modems can accept a much wider set of commands than these because it is through the command set that all the protocols are actually set. The purpose of good communications software is to provide a user-friendly, English-language control mode to send these other commands to the modem for the user. If the communications software does not do this, then the user must use the modem's command set (found in all modem user's manuals) to send the specific command strings for baud rate, bits per character, parity, stop bits, and any other necessary modem setup parameters that are needed for effective telecommunications.

Data Mode

When telephone contact is successfully accomplished with the "dial" command, the modem automatically switches from command to data mode. In data mode, everything that is typed at the keyboard is interpreted as data to be sent to the host computer. This is when the actual process of telecommunications takes place. The other modes which have been discussed are designed to set up the hardware and software for data transfer. In data mode, the user's computer is in direct contact with the host system. The bulletin board system, information utility, or mainframe operating system being run by the host computer now recognizes the user's computer as a remote terminal on its system and behaves as though the user's computer is directly wired to it. All menus, prompts, and information that the user sees on his or her computer screen are being sent from the host computer. The host system is in control of the online session. The host computer recognizes all the keyboard characters that the user types, including the "escape," "delete," "break," and "control" keys. For this reason, while in data mode, certain non-standard key combinations are usually necessary should the user wish to regain control of her or his computer. Check the communications software's user manual for this information. For Hayes compatible command sets, pressing the "+" key three times in rapid succession returns the modem to command mode. Pressing the open-apple and escape keys simultaneously (Apple Access II) or Alt and Z keys (Procom Plus) puts the program back into control mode. The ability to switch modes while online is one of the advantages of communications programs. It permits access to utilities and disk operations while still online, as well as the ability to regain control of the modem should the host computer experience a system "crash" and become unresponsive.

Asynchronous and Synchronous Communications Systems

Two terms often come up in discussions or articles on telecommunications: asynchronous and synchronous. These terms refer to the way the data is organized for transmission and reception.

Asynchronous telecommunications means that either modem can send data to the other whenever it wishes. This is the type of data organization used by most personal telecommunications systems. In fact, special hardware must be obtained to use synchronous technology. In asynchronous systems, data exchange is not controlled by a master clock. Each modem has its own built-in oscillator that functions as an independent quartz clock. Both modems operate independently and follow a built-in program that loops over and over, looking for a start bit signal to appear. When the start bit is detected, the modem switches to a data collection routine and uses its own internal clock to tell when to determine the binary status of each incoming bit. The bits are assembled into sets until the stop bit is received. Then the data byte is passed to the computer and the modem goes back to looping through its wait cycle, looking for the next start bit. The trade-off for being able to send data whenever it is convenient is the need for extra control bits to start and stop the process of assembling sets of bits into meaningful bytes of data. This increases the size of each character by 2 or more bits. This is an increase of 20% or more, slowing down data transfer by at least that amount and adding additional sources of error.

Synchronous telecommunications systems, on the other hand, use a single clock to control both modems. As a result, the modems are always in an identical operating condition and it is not necessary to send start and stop coordinating signals as part of the data, making the process of transfer much faster and more accurate. These systems are expensive and are generally used by government and industry.

Asynchronous and synchronous are not protocol settings controllable with software or modem commands. They are special circuits and internal programs that are in the hardware. Personal telecommunications systems only use asynchronous technology, and personal computer stores and dealers only carry and advertise asynchronous equipment.

COMMUNICATIONS SOFTWARE FEATURES

Communications software programs are intended to simplify the telecommunications process by managing most of the technical aspects of using a modem. These programs, sometimes known as terminal emulators, come with a variety of features ranging from the most basic to the very sophisticated. When selecting a communications program, consider the extent of its basic, convenient, and advanced features.

Basic Communications Program Features

Do not consider using any communications program that does not provide easy-to-use computer control, modem command, and break key capabilities (such as Control C in BASIC). All communications programs offer terminal mode. For the simplest programs this is the extent of their capabilities. This means the user must memorize the computer control, modem command, and break key instructions. Furthermore, if any modem or computer protocols cannot be preset, they have to be reissued each time the terminal program is run.

Such simple communications programs perform the required telecommunications function, but a lot of effort is required to manage and control the data being exchanged. A minimally adequate communications program should provide both control and terminal modes, preferably with a main menu to manage the system. The control mode should offer a means of presetting and editing modem protocols for baud, bits per character, parity, stop bits, and duplex that are automatically sent to the modem when the program is started up. It should also offer an online printing option as a means of obtaining a permanent record of the online sessions.

Convenient Communications Program Features

The use of a communications program that offers a complete array of convenient modem command and computer control features can contribute significantly to the ease of telecommunications. These features make use of the computer's data handling capabilities to create what is generally referred to as a "user-friendly" program.

Menus and Help Screens

The use of menus and help screens are typical features of such programs. Menu-driven programs have become the hallmark of user friendliness. Menus guide the user through the selection of communications software operations. They specify the choices available at each step of the program and eliminate the need to memorize instructions or refer to manuals.

Help screens are monitor displays that explain concepts and program operations to the user when she or he is having problems. The extent and usefulness of help screens depend on the amount and kind of information the author of the program has included.

Installation Options

With so many combinations of interfaces, modems, and printers available for each brand of personal computer, the risk of incompatibility between hardware devices is high. A communications program that can work with many different pieces of equipment easily supports changes and upgrades in the telecommunications system. These communications programs contain routines that anticipate certain combinations of equipment and permit specification of the interface slot or port, interface card and modem type, and printer slot and type. The better programs also provide a null modem setting should some of the hardware not be specifically supported by the program. This setting allows anything typed at the keyboard to pass straight through to the modem. The user then uses the modem's command set to control the telecommunications setup. This means that the program can work with many hardware configurations or allow the user to control the modem.

Additional features that add to the ease of operation are the ability to save specific installation settings as separate files, to edit installation settings, and to send special command strings to the modem when the system is started. Saving separate installation settings can be convenient when using several different host computers or hardware configurations. The editing function supports changing and upgrading the system without having to buy new software. The command string feature permits use of special or unique functions that a modem might possess that were not specifically anticipated by the communications software programmers.

Protocol Selection

The ability to set baud rate, bits per character, parity and stop bits is essential, but a number of other features contribute to the smooth and meaningful exchange of information. The ability to set terminal type, line feeds, Xon/Xoff, and video and screen formats make a communications program much more useful. Together these features make for easy protocol management.

Other useful management features include character echo option, modem status indication, meta-character and character substitution support, and a hang-up option. The character echo option causes the communications program to print (or echo) characters typed by the user to the screen. In full duplex operation, the characters appear on the screen because the host system has echoed them back. In half duplex, the program should automatically echo the characters to the screen. It is useful to be able to disable the echo feature to prevent people who may be watching from seeing what is sent. This applies primarily to passwords and other log-on codes that are not generally public information, but also permits private communications.

Modem status, either on screen or via window commands, permits rapid checking and editing of the protocol settings. This can be very convenient if the protocol settings are a part of the installation process and are stored as part of a configuration file.

Meta-characters are "keys" that the computer keyboard does not actually have, but that can be simulated by pressing special combinations of other keys. This can be important when communicating with a host system that uses a different computer and keyboard.

Character substitution is very useful for avoiding conflicts between the host system's use of certain non-standard keys and the personal computer's response to them. The most common need for character substitution involves the "delete" or "rub out" key. No standard command exists for "back space and erase" (also called the *destructive backspace*), and many different keys including the back arrow, delete, clear, break, and escape are sometimes used. It is very useful to be able to instruct your computer to translate the ASCII code that it uses to mean "back space and erase" to the code that the host system is using. Once properly set, your keyboard may be used in the usual way and the software takes care of correctly communicating the command to the host system.

A hang-up or "panic button" option is useful for dealing with unexpected situations such as a "crash" of the host system that causes it to become unresponsive. A crash of a communications program often does not cause a disconnect signal to be sent to the modem, so it continues to maintain the connection. The hang-up feature permits the user to manually break the connection by instructing his or her modem to hang up the telephone.

Auto disconnect is a feature that responds to the modem's dialing condition, automatically hanging up the telephone line and returning control to the communications program when certain signals occur. Among these conditions are *delay times* before abandoning a telephone call that is not answered, detecting a *busy signal*, and sensing the loss of the other modem's *carrier signal*. In each of these cases, communication is either not established or lost, so the telephone call ends automatically and program control is returned to the user.

Autodial

Many modems contain autodialing circuits that can activate the telephone company's switching network and place a telephone call automatically. Generally, autodialers can generate both keypad tones and rotary dial pulses. If the modem has these capabilities, it is convenient to be able to have the communications program place the calls.

There are several different implementations of the autodial feature. The simplest method permits the user to type the telephone number at the keyboard and have the modem place the call. This is sometimes called the "quick dial" option. A more full-featured program maintains a directory of telephone numbers that are called frequently, and places a call by choosing the desired number from the directory. Even more convenient are programs that permit automatic redialing of numbers if the connection is not established the first time. Autoredial can be either manual or automatic, single-try or continuous, until the connection is made or the process is interrupted.

Capture Buffer

A capture buffer is a useful feature for high-speed recording of the telecommunications session. It generally provides at least the ability to save the captured information to disk. Even more useful is the ability of some programs to inspect and/or edit the recorded information before saving. The option to print it as well is useful. A "control character filter" feature is helpful in these circumstances for removing embedded control characters from the ASCII text. These characters are needed for telecommunications but can often cause unexpected behavior in printers and disk drives.

ASCII File Save

Most communications programs that offer a capture buffer also permit saving the recorded ASCII files to disk. The variation among these programs includes whether this save feature is automatic or under manual control. Manual control usually allows specifying automatic operation, or permits inspecting the recorded file before deciding on the save. If editing the capture buffer is available, the disk save feature is generally manual.

ASCII File Send

The ability to automatically send an ASCII text file is useful for saving online time and charges by creating and editing files prior to the online session. This feature is also useful if the same message is sent to more than one person or bulletin board system. To make the feature versatile, some sending parameters need to be set either at the time of transmission or during modem installation or protocol setting. These parameters include character and line delays, and response to prompts.

The character and line delay times are necessary because most personal telecommunications systems assume that the user is actually sitting at the keyboard typing the messages. The computer is capable of sending text three to ten times faster than a person can type. Many host systems cannot handle characters that quickly. Good default settings for most systems are 30 milliseconds between characters and 100 milliseconds between lines. In full duplex operation, watch the text as it is echoed back to the screen. If words are missing characters, or lines are missing all or parts of words, then stop and increase the delay times.

Some host telecommunications systems are line oriented rather than character oriented. This means that they do not process the received characters until a carriage return is sent. These systems usually can be identified through their use of a prompt character that signals the system is ready to receive the next line of text. Most ASCII file sending programs can be set to line mode. The user is asked to specify the prompt character that the program should wait for between lines. A word of caution when using such a system for ASCII file transfer: if the prompt character is not received correctly by the program for any reason, it waits indefinitely to send the next line. In this situation, break out of the ASCII send feature and try again.

Disk File Management

With the ability to save captured files to disk, it is useful to have a built-in utility to manage such files. This utility should permit formatting blank disks, cataloging the contents of a disk and showing how much space is used and available, creating files (and subdirectories), and deleting files from the disk.

Chat Mode

It is not uncommon for a telecommunications user to pause in the middle of a voice telephone call to exchange files with another person. The ability to go online with the telecommunications system when the telephone line is already connected is known as chat mode. In this mode, the person does not need or wish to dial a number. One of the parties must be able to activate his or her modem so that it sends out a carrier signal. This requires a modem that can either be set to answer mode or that automatically switches to it when online.

A chat is established by having one or the other modem send out the answer mode carrier signal. When this is heard on the telephone line by the other party, he or she activates his or her modem in the originate mode. Assuming that both modems have been set to the same protocols (usually arranged by voice before attempting the chat), they establish a telecommunications link. When this has been accomplished, both parties hang up their telephone handsets and whatever is typed on one computer keyboard appears on the other. Files can be sent and received. When the telecommunications have been completed, both parties pick up their telephone handsets and direct their modems to hang up. When the modems hang up the carrier signals disappear and voice communications may continue.

Advanced Communications Program Features

Some telecommunications programs support a variety of very sophisticated features. These features include improved adapatibility to hardware and advanced operating functions not necessary for most occasional telecommunications. If, however, telecommunications is used in a business or on a very regular basis, some advanced features may be desirable.

Support for hardware options includes the ability of the program to detect and use memory expansion cards, RAM disks, hard disks, and clock cards. These hardware devices enhance the capabilities of personal computers by giving them larger, faster memories and the ability to identify and use real-time information. When using a computer with any of these enhancements, have them available for use when telecommunicating by making sure that the communications program specifically supports this equipment. This hardware adds to the number of possible installation configurations and increases the chance of incompatibility. If these features are essential, see a working demonstration of the complex hardware and software system before buying anything.

Software enhancements are many and are beyond the scope of this book to deal with in great detail. The most useful features, however, can be discussed in principle. Terminal emulation is probably the most common advanced software feature. It allows a host system to see a calling computer as just another of its terminals. Among the two most common terminal types that host systems expect to see are the VT52 and the ANSI VT100. Many communications programs can emulate the behavior of these terminal types, permitting the user to take advantage of their special features.

It is sometimes useful to be able to determine the status of a program and monitor at a glance. It can also be helpful to know the time, date, duration of connection, and memory capacities. Many advanced communications programs display an optional *status bar* that shows this type of information.

The ability to save autodial parameter files is another common advanced feature. Autodial parameter files contain information beyond simple telephone numbers. These files, sometimes called command or macro files, can contain extended dialing codes needed to activate local switchboards and wait for special dial tones. It is frequently possible to include autodial chaining information that instructs the program to call another telephone number or use another autodial parameter file when the current call is ended.

Autodial can contain log-on macros that are instructions to the program about host system prompts and automatic responses to them. Many macro commands can be "learned" by the program itself by having it record a manual log-on session. After that, it can repeat the sequences automatically. Macros can often be created manually either within the program or with a word processor. Programs that support macros generally outline the macro programming language commands in their manuals. Advanced users can instruct communications programs to operate virtually independently with the proper macro command files.

Window-based command options is a recent feature that greatly facilitates control of the telecommunications process. Windowing uses special memory areas to save data from the screen when control menus are displayed. In this way, once the command choice has been made, the screen can be returned to its former state. In programs without this feature, the appearance of command menus or the typing of command strings corrupt the contents of the monitor, perhaps even causing data to scroll off the screen.

A sophisticated version of windowing permits reviewing any portion of the data that has been received, even if it has scrolled off screen, and then return to the screen. In this fashion, information can be retrieved, perhaps for names, numbers, dates, or addresses. This feature is sometimes called "scrollback." Along with scrollback, it is handy to be able to edit or save certain parts for future reference, so this decision does not need to be made as data is being received.

With the trend towards integrated software packages, it is not uncommon to find communications programs with built-in word processors. With frequent ASCII file transfer, this can be a very efficient option. A built-in word processor avoids the need to convert files from one type to another and guarantees disk operating system compatibility.

Sharing data from host systems is done much more easily with programs that integrate spreadsheets and databases. This level of program integration is not common yet, but the trend is clearly in this direction. One of the great potentials of telecommunications is found in its promise of quick and efficient exchange of digital data ready for analysis and processing by other types of programs.

Several advanced file handling features can be essential when handling large files or using high baud rates. If the files being exchanged are likely to exceed disk capacity, the ability to save "segmented" or divided files is critical. To transfer files at high baud rates, or print one file while receiving another, a printer buffer is needed. This is an area of memory that holds data until the printer can print it. It allows data to be received at a faster rate than it is printed (up to the limit of the buffer's memory size) or permits part of the computer to hold and print data that is different than what is presently being exchanged.

The Xmodem, Ymodem, or Kermit error-free transfer protocols are necessary to exchange program code or binary data frequently. These protocols provide error-free exchange of 8-bit binary data and ensure the integrity of programs and information. For very sophisticated file transfer operations, the communications program should function in batch or disk mode as well. In batch mode, several files can be transferred at once without manual intervention. In disk mode, all the data on a given disk can be exchanged—sort of a "disk copy" by telecommunications.

Finally, it is possible to turn the computer into a host system that can be called by another computer. The main purpose of this advanced feature is the ability to access files from a distant location. This program feature requires a modem able to operate in autoanswer mode. Be aware that any computer could call into the system, so be sure that the password protection system is adequate to correctly screen out unauthorized users. Obtaining complete access to a computer's files, regardless of location, can be a very powerful and efficient feature. It might well be a better solution than carrying around a set of disks.

9

On the Horizon: Promising Technologies and Methodologies

ADVANCES IN TELECOMMUNICATIONS TECHNOLOGY

Technological advances in the computer, communications, and video fields are leading to a convergence of systems as the information age progresses. The ability to create, store, process, manage, and share information is accelerating as the ability to combine computers with video and link both with telecommunications grows. As information technologies become more prevalent in everyday life, they become increasingly useful to teaching and learning. This chapter reviews recent technological developments and some innovative educational projects presently under way.

Telecommunications includes "all services, products, media and methodologies used to deliver information electronically, from a simple telephone to sophisticated fiber-optic networks" (Charp & Hines, 1988, p. 94). It is the network concept, however, that is central to the rapid development of information technologies. A network is nothing more than the interconnection of computer, video, and communications systems. Once connected, the resulting telecommunications system becomes more than the sum of its parts. Its capabilities often exceed those of the component systems acting separately, and the system itself becomes a unit that can be merged with other component parts. The process of development becomes a self-sustaining cycle of new systems, followed by new uses, followed by yet newer systems.

Much of the technical development taking place today deals with solving the problems of interconnection among different types of information devices. This chapter examines some of the ideas about networking standards that are beginning to emerge. Educational initiatives often result from advancing technical capabilities, yet educational applications are rarely considered by the creators of networks when designing their systems. A great deal of excitement exists in education today in the search for new ways to teach and learn using the increasingly easy access to information provided by electronic networks.

As efforts progress towards wider, simpler electronic information resource networks, a common concept of what is involved in telecommunications is emerging. An important idea is that of the *channel:* a physical or logical pathway over which information can be transmitted (Charp & Hines, 1988). The channel concept includes circuits, lines, access links and facilities, and may be realized with wire or fiber optic cable, and/or radio waves such as shortwave or microwave. Telecommunications channels can be either public or private, and the physical limitation of close proximity between two information devices can be eliminated if both can access the channel.

Channels can be either switched or non-switched. The public telephone network is an example of a switched system. By dialing a number, the telephone company switching system can be instructed to make connection with a particular unique location practically anywhere in the world. The purpose of the switch is to "establish and maintain a physical connection between the sender and receiver. . . . Non-switched systems are dedicated physical systems for a predetermined set of users" (Charp & Hines, 1988, p. 95). Non-switched systems are typical of local computer networks used to link workstations within an office, company, university, school, or municipality. They are inherently private, although it is possible to provide links to public switching systems (i.e., through a modem to the telephone system).

Normally, when a user activates a telecommunications channel, the channel is unavailable to others until the user completes the online session. For example, when a computer calls a local bulletin board, the telephone cannot be used for other incoming or outgoing calls—it's busy. This is frequently not an efficient use of the channel's resources. Telecommunications occurs electronically at the speed of light: bits of information travel at a speed of 186,000 miles per second. Hundreds of thousands of bits of information can be exchanged in milliseconds. It is a waste of valuable telecommunications resources for a communications channel to be tied up by a single user, since most of the time the channel is idle while the system waits for the next character to be sent. In order to make better use of telecommunications channels, efforts are under way to perfect what is called packet switching.

Packet switching permits multiple users to be on any given channel at the same time. Some number of bytes of information (usually 128) to be sent from one user to another is collected into a single bundle and combined with an "address" header to form an information packet. Each telecommunications system on the channel works as both a "mail box" and a "post office." As each packet is received, a system checks to see if it is addressed to itself. If so, it saves the packet for use by that system; if not, it sends it along to the next address. In this fashion, all telecommunications systems on the channel (a telephone line, for example) are active at all times, but they are not busy in the conventional sense, responding only to incoming packets with their specific address. At present, most packet switching developments are found in local area networks (LANs) where each device on the system needs to be active at all times.

The public telephone system is not yet ready to support widespread packet switching. For one thing, the system was not designed to have all telephones "off the hook and active" at the same time. For another, a fair and effective tariff rate structure for theoretically continuous local and long distance access has not been developed.

Radio Telecommunications

One example of public access packet switching is the educational packet switching project using an amateur radio packet network access station set up at the Technical Education Research Centers (TERC) in Cambridge, Massachusetts, by licensed radio amateur Bruce Seiger. Created to explore the possibilities and problems of using shortwave radio as a communications channel, the system has established a relationship with the packet bulletin board system (PBBS) operating on a frequency of 145.05 MHz with the radio call sign N1BGG in Charlestown, Massachusetts, and with COSIN (Computer Student Information Network) operating under the amateur radio call sign WB2MIC on the same frequency in Wells, Vermont (Seiger, 1988). Amateur radio packet bulletin boards are similar to their telephone counterparts. They can be used for electronic mail, posting bulletins, and engaging in extended teleconferences. Operating in their "post office" mode, radio packets can be sent virtually anywhere in the world.

A requirement for operating an amateur radio packet station is a Technician Class amateur radio license. The license is earned by passing a written exam on basic radio theory and sending Morse code at 5 words per minute. This may sound formidable, but 40 third-graders at the Eastham Elementary School in Eastham, Massachusetts, were able to get their licenses in order to participate. In any case, any appropriately licensed amateur can oversee the operation of a packet station. Many amateurs are willing to volunteer their time to help a class explore the world of shortwave packet radio. More information on radio license classes and exams, and the names and addresses of local amateurs who can be contacted, can be obtained from the American Radio Relay League (ARRL). See the Resource Section for more information.

Avoiding the expense and time involved in using the telephone system for telecommunications prompted the packet radio project at TERC. With packet radio, the connect time is free and the user can call a PBBS at anytime. During peak hours there may be some delay in sending messages, but there is never a busy signal. The primary monetary requirement is the investment in a radio transceiver ($300), terminal node controller computer interface, similar to a modem ($100), and an antenna ($30).

The goal of the radio packet system operated by TERC is the active promotion of networking for cooperative scientific inquiry. Students and teachers from classrooms around the country use the radio packet system to send science-related questions to TERC and its science consultants. The responses are then sent back over the packet network. Jim Packet, whose class at the Eastham Elementary School participated in the TERC project, couldn't be more enthusiastic.

> I've been teaching for over 19 years and let me tell you, every teacher needs something like this every 19 years. It was like starting fresh all over. The kids loved it; they ate it up! They couldn't wait to get messages from Minot, North Dakota, or Walnut Creek, California. They didn't even know Walnut Creek existed before this.
>
> It was the greatest learning motivator I've ever seen in all my years in the classroom. And this year's class cannot wait to get started. They've heard all about it from last year's class, and are ready to do it. That's the kind of curriculum you strive for. (Packet, 1988)

The amateur packet radio system that TERC has established is an operational example of what Donald Stoner proposes as a Public Digital Radio Service (PDRS). Stoner filed a petition with the Federal Communications Commission in 1986 to set aside a portion of the 52-54 MHz amateur radio band for a nationwide PDRS telecommunications channel. As he envisions the system, individuals purchase specially built "smart transceivers" (combination radio transmitters and receivers with built in modems) which they attach to their computers and which act as access devices to the radio network. Much as amateur packet stations operate today, the "smart transceiver" sends out a signal requesting all nearby packet stations to reply. It determines the ten stations nearest to itself and forms a local area radio network with them. Thereafter, it continuously monitors the radio frequency for messages. When such messages are heard, it determines if they are addressed to itself or are being routed through it for delivery to another nearby station. As the number of users increases, the potential for worldwide radio packet telecommunications grows. Stoner imagines that eventually anyone could have access to unlimited telecommunications capabilities for the cost of the "smart transceiver" (Stoner, 1986).

Another experimental program utilizing radio communications to create a telecommunications channel to alleviate long-distance telephone charges is being run by the Center for Mathematics, Science and Environmental Education at Western Kentucky University (Office of Technology Assessment, 1987). This program makes use of the Early Warning System microwave network to broadcast computer courseware to rural schools. Twenty-one schools in 14 districts are connected to the microwave relay stations through local telephone lines thereby establishing a network with the university's mainframe computer.

The use of commercial FM radio and television channels for the broadcasting of digital information is another development currently being explored in several places around the country. The Software Communications Service, an organization of 17 state public broadcasting systems and 5 Canadian provinces, is experimenting with the ability of television stations to distribute instructional software to thousands of classrooms around the nation at a fraction of the cost of conventional dissemination methods (Office of Technology Assessment, 1987). The digital information is sent out along with the normal picture and sound signals as a specially coded "subcarrier" that is not detectable to the general audience, in much the same way that Muzak is sent to stores over FM radio. By using a special receiver, a school can pick up the subcarrier signal and demodulate the digital information. The limitation on this telecommunications channel is its one-way broadcast nature. At the present time, private users are not permitted to transmit signals in the radio and TV broadcast frequency bands.

"Sofcasting" is a variation on the theme of using commercial broadcast signals for the distribution of digital data (Joyce, 1985). A creation of Don Stoner, originator of the PDRS concept, sofcasting is the broadcast of digital signals as part of regularly scheduled commercial radio programming. Produced by the National Digital Network, the residents of Seattle, Washington, can now receive public domain software and other digital data by listening to AM radio station KAMT's "Download" program at 9:30 PM Sunday evenings. The radio program's 30-minute format is similar to many other special interest variety shows, and includes guest interviews, product reviews, and industry news. What makes it unique is its inclusion of 10 to 30 seconds of digital data transmission across the air. AM stations transmit the data at 2400 baud, while FM stations can use 4800 baud. To the average listener, this sounds like a brief period of static or hum. For the listener equipped with a computer and decoder, it contains computer data or a public domain program. The decoder is a $70 unit called a Shuttle Communicator, invented by Don Stoner and produced and marketed by the Microperipheral Corporation, of which Stoner is Vice President of Engineering. Microperipheral also produces the communications software that allows the Shuttle Communicator to work on almost all personal computers. The data sent to listeners is free; advertising revenue is used to cover the costs that are involved. The National Digital Network is presently negotiating with radio networks and cable systems around the country interested in distributing the radio program, including the operators of the Westar II satellite. The limitation of sofcasting is the same as that for broadcast subcarriers—it is a one-way operation. Receiving stations cannot send information to the originating broadcast station (Sofcast Corporation, n.d.; "Computer Data," 1984).

Communications Satellites

Communications satellites offer another possibility for widespread telecommunications. As the direct broadcast satellite technology improves, driven by the interest in home entertainment satellite reception, the size and complexity of receiving "earth stations" has been greatly reduced. Direct reception of communications satellite signals can now be obtained for about $2000 using an 8-foot diameter receiving antenna. This is well within the budgetary abilities of many educational institutions. One educational satellite communications project presently in operation is the Australaskan Writing Project, involving 51 schools in Alaska and 53 schools in Australia. Twice a week the schools pair off to allow students to discuss areas of interest. The Alaskan students use the AlaskaNet telecommunications system and the Australians use the MIDAS data network. The information is sent via Alascom's Aurora satellite through the AlaskaNet/Tymnet gateway in Seattle to Tymnet's undersea cable connection between San Francisco, California, and Sydney, Australia. Following a six-phase curriculum plan, students share their writing with peers halfway around the world. Project creator Malcolm Beazley, Australia, summarizes the project in the following way:

> We have had a modest beginning and have evolved over a period of time into a rather exciting enterprise which provides a creative outlet for children while teaching them about the English language and writing in particular. We all feel that this could be considered the beginning of a sound foundation for major international understanding. (Smirnoff, 1987, p. 9)

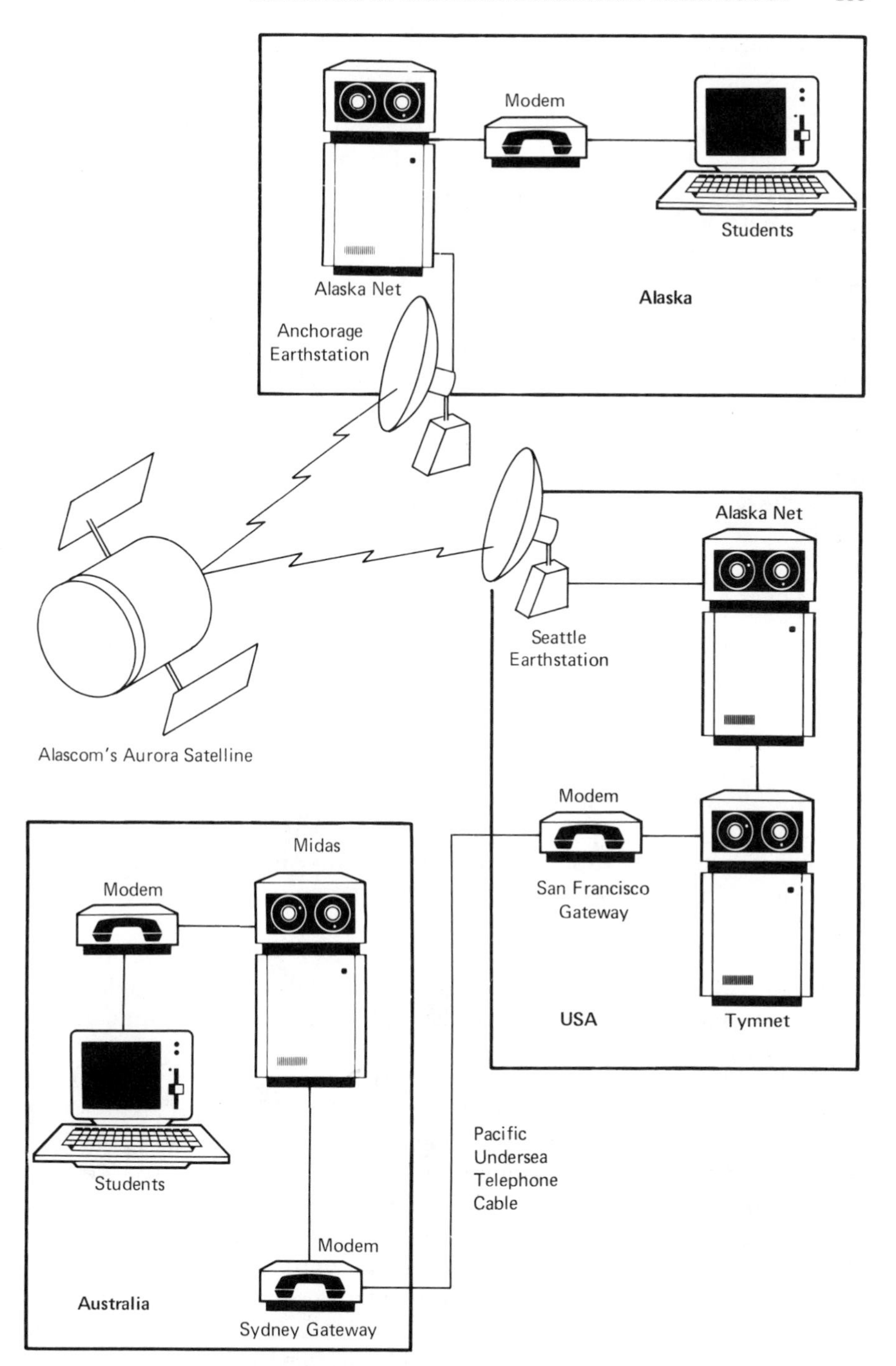

FIGURE 9.1. Australaskan System

Fiber Optics

Another rapidly developing technology with important implications for telecommunications is the use of fiber optic cables in place of copper wire. Activated with light from miniature lasers, fiber optics can carry billions of bits of data in cables no bigger than the diameter of a pencil. The rate of data transmission depends on the available bandwidth. Larger bandwidths can send data faster, but for fewer users. Due to their inherently greater channel capacity, fiber optics cables provide greater bandwidths without eliminating users. Participating in a joint project with Harvard University, the NYNEX Service Company will link four of the University's computer centers using fiber optic cables that can handle data exchange at 10 million bits per second, or about 178 times faster than present copper wire technology. Known as the Metropolitan Area Network (MAN) project, NYNEX and Harvard are jointly experimenting with network design and communications protocols (New England Telephone, n.d.).

Local Area Networks

A Local Area Network (LAN) is actually a small, non-switched telecommunications channel designed to support the needs of a defined group of users. Dedicated, non-switched communications networks linking terminals with mainframe or mini computers have been common for many years. Recent advances in the development of standard data transfer protocols, however, have moved such systems in the direction of becoming miniature telecommunications networks. The data transfer standards are being developed as a response to the need to link larger numbers of digital devices, made by various manufacturers, to common data sources.

Structurally, a local area network is a system that links two or more computers together so that they can share data files or other devices. Communication among computers and peripherals consists of data packets transmitted along some form of channel, usually a cable. Four configurations are common: star, tree, ring, and bus.

In a star configuration, one computer is located at the center of the system, attached to every other device in the LAN with dedicated lines that radiate out from it (see Figure 9.2). The central computer acts as the "file server" or master system. It communicates with each of the other devices to determine and process their data requests. For example, two computers on the LAN might wish to use a particular program. The file server fetches the program from a hard disk drive memory and loads it into the two requesting computers. If another computer wishes to print a document, the file server accepts it over the LAN and sends it to a laser printer. Still another might wish to save data, in which case the file server accepts and stores the data on the hard disk. Basically, the file server is in charge of

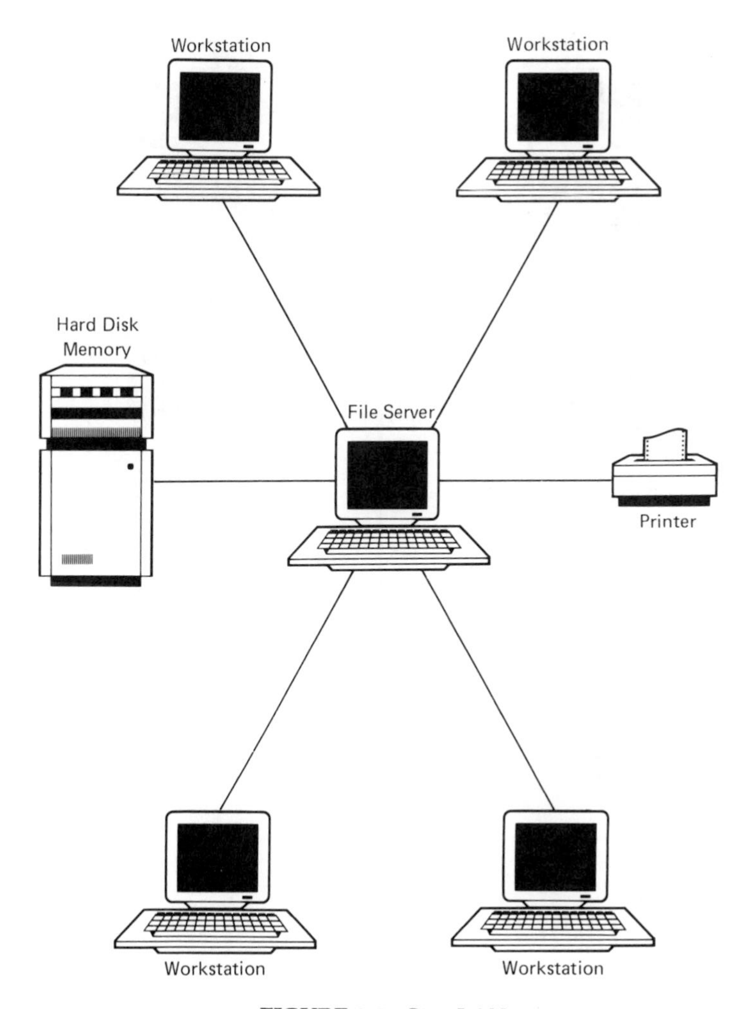

FIGURE 9.2. Star LAN

transferring programs, data, and text in an orderly fashion between the devices in the LAN.

With the tree structure (see Figure 9.3), the file server acts as the base, or root, of a tree whose trunk is attached to all the individual devices in the LAN by way of specific branches. The operation of the LAN achieves the same purposes as the star layout, but the communications occur over a shared channel so each device needs to recognize and send "handshaking" signals to maintain orderly data flow.

A ring LAN connects all the devices together in a continuous loop (see Figure 9.4). One of the devices in the loop acts as the file server, but each device in the ring must be "smart" enough to recognize data packets

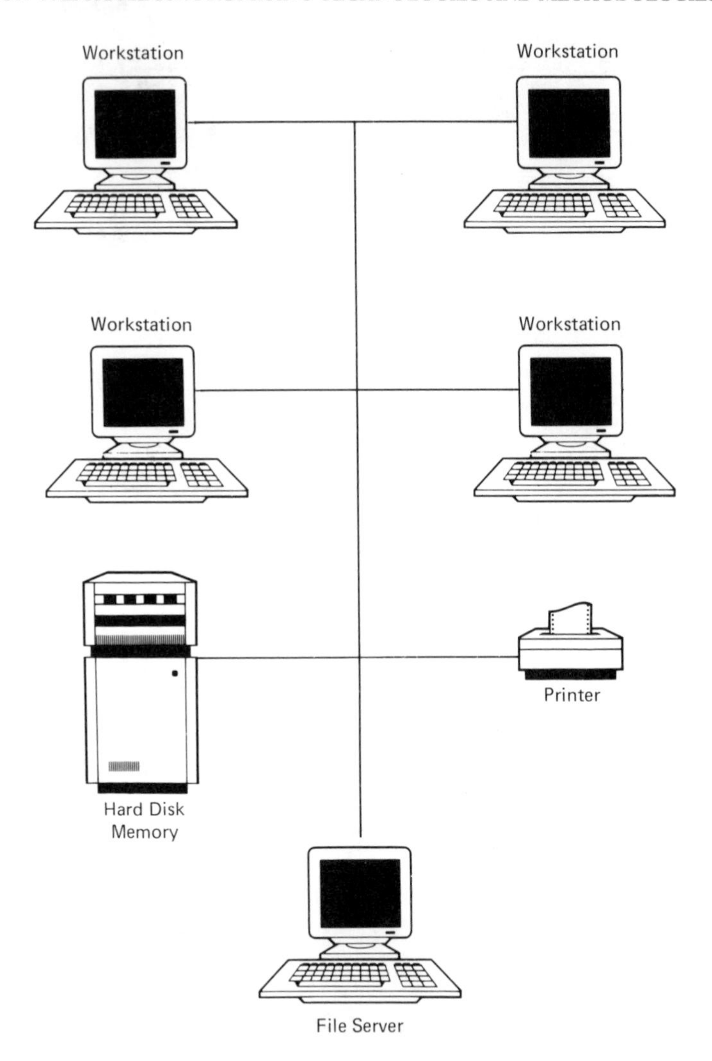

FIGURE 9.3. Tree LAN

directed to itself. This type of LAN accomplishes the same things as the star or tree configuration, but data packets flow around the ring until they reach the correct device. The communications interface for a ring must therefore be capable of examining each data packet to determine whether to accept it or send it on.

A bus type LAN consists of a single cable open on both ends to which network devices are attached (see Figure 9.5). The bus channel serves as a "backbone" to which all the other devices are attached. All that is necessary to connect a digital device to the bus is a "tap" connector. When

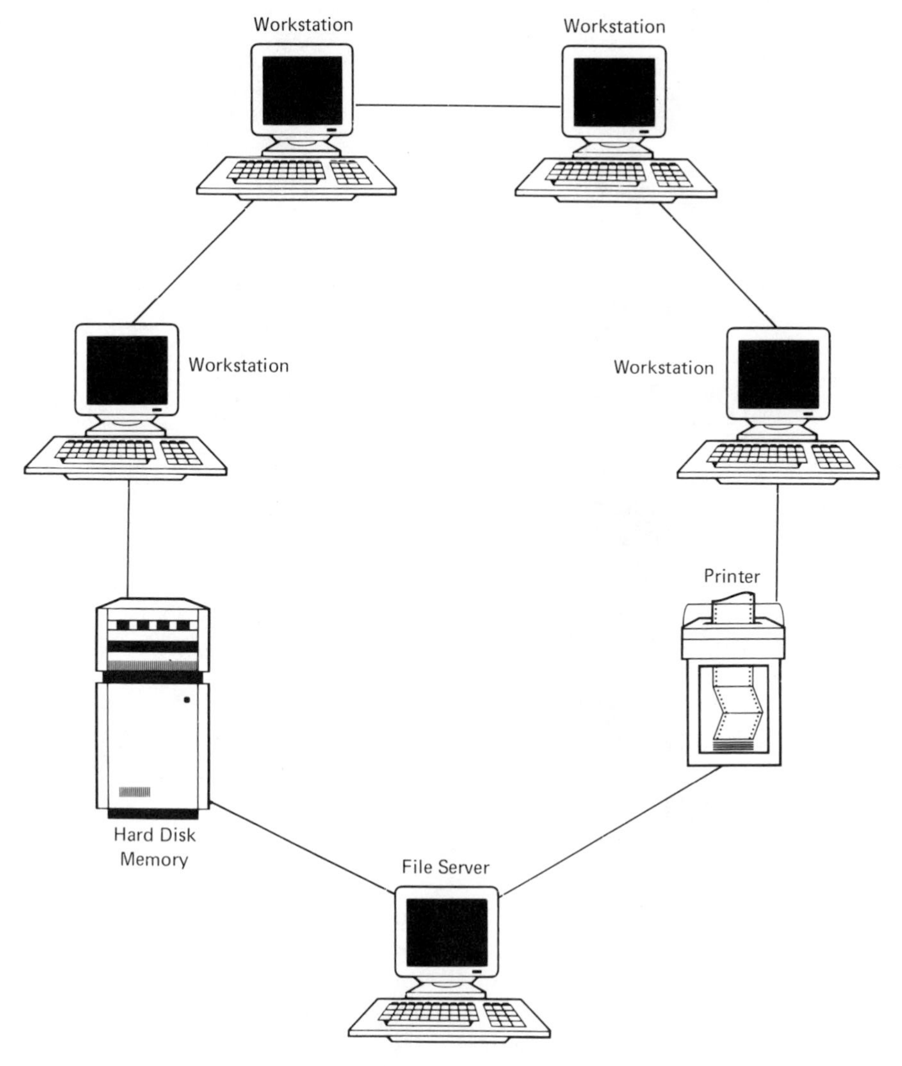

FIGURE 9.4. Ring LAN

information is put on the bus, it travels away from the source connection in both directions and is received by all the devices on the system at the same time. Like the ring LAN, therefore, all the devices must have "smart" interfaces to recognize those data packets directed to them.

One of the primary purposes for installing a LAN is to allow several computer workstations to share large memory storage devices and expensive printers. Schools have often moved toward the use of LANs as a cost-effective way to manage a large software collection, providing access to many programs without the problems associated with individual floppy

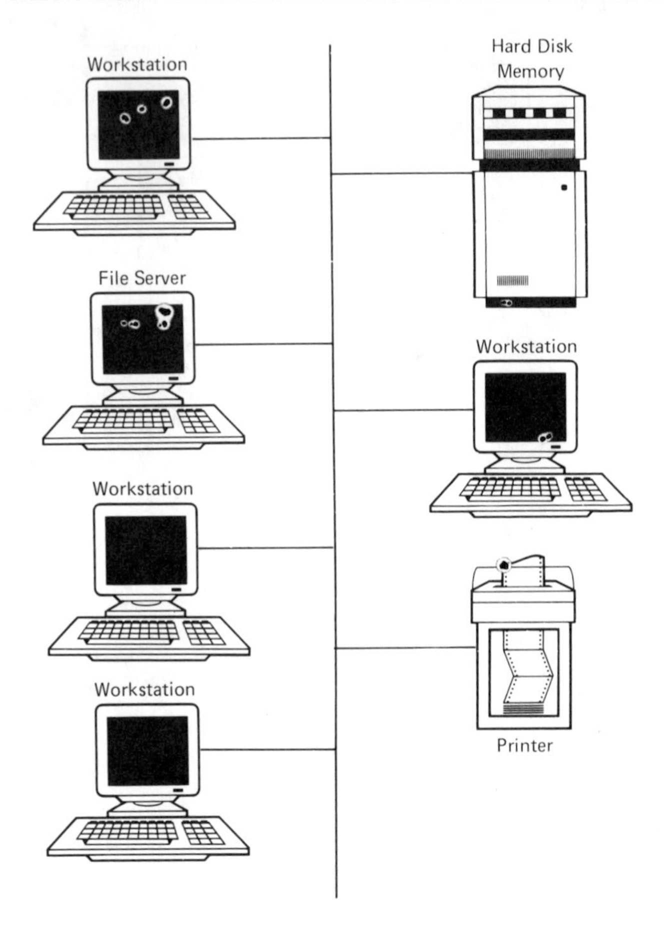

FIGURE 9.5. Bus LAN

disks. Many software producers are inclined to negotiate site licenses with LAN users because of the inherent security in the hardware-based sharing of multiple copies of their programs.

In order to function within a LAN, each computer and peripheral device must have an interface card connected to the LAN, and a network program that can communicate through the interface. This is similar to telecommunications over phone lines using modems connected to RS-232C ports. Hence, the implications for the convergence of LAN technology with conventional telecommunications are immense. In principle, anything that can be done with telecommunications over phone lines can be done over a LAN. It is this very similarity that is leading to a convergence of the two technologies. A number of educational projects already have been designed around the LAN concept as a means of deriving the benefits of telecommunication's collaborative possibilities without the financial burdens of telephone system or information utility access charges.

Two such projects are Dr. Trent Batson's at Gallaudet University and another at Carnegie Mellon University. In both instances, microcomputer LANs are used to conduct class discussion and manage interactive assignments, at Gallaudet to teach English to deaf students and at Carnegie Mellon to teach foreign language. The real-time communication process creates an authentic language-based environment within which students are immersed. The system consists of IBM PCs connected via either Fox Research's 10Net or a Novell network running RealTime Writer network software. The key feature of this network software is its ability to permit any user on the network to type a message and send it to all the others (Johnson, 1988).

Another popular LAN network found in the public schools is the Corvus Omninet with Constellation III network software. This network is compatible with the popular Apple II family of computers widely used in elementary and middle schools. The Bank Street College Earth Lab project is designed around the Corvus LAN system. Directed by Dennis Newman, the project engages sixth-grade students in Manhattan in the study and analysis of weather-related science principles. Observations, hypotheses, and reports are shared on a network of 25 Apple IIe computers. Students use an integrated package of computer programs and tools to compile and share data and to write reports. The purpose of the project is to create a model for using computers to teach scientific collaboration in schools. The primary pedagogical approach is to use cooperative learning among students to improve their learning of science content, as well as their understanding of the social (cooperative) processes of the scientific community ("Earth Lab," 1987; Newman, 1986).

Sharing on the Earth Lab LAN is facilitated through the use of common files to which all users have access. Designed around Apple Computer's ProDOS hierarchical file structure that supports hard disk drives, the Constellation III network is controlled by the Earth Lab Interface operating shell that manages student accounts and permits each student to find his or her own personal subdirectory of files, including private files as well as tool programs for data analysis and reporting and public files for sharing information.

The emerging standard for local area network interface connections is called the Integrated Services Digital Network (ISDN) (Charp & Hines, 1988). This proposed standard would establish an international communications network architecture and set of data exchange standards that would be embedded in all LAN systems as well as in the telephone exchange system. As such, it would become possible for any digital data device to communicate with another by any available channel. The concept for this proposed standard is to translate data specific to any particular digital device into a common data interchange format, sort of like ASCII only much more comprehensive and sophisticated. Eventually, it should be possible to connect a computer to any telecommunications channel and communicate with all other devices online without thinking about the software or hardware being used. The ISDN system's embedded "intelligence" takes care of the details.

The ISDN standard is comprised of seven levels of services and protocols. Each represents a "layer" of intelligence that lies between the user and the telecommunications channel. The combination of levels permits all forms of digital hardware to be adapted to work on the network system. The first layer is the *physical layer*, providing the direct electrical interface to the network. Included in the physical design is the *data link layer*, providing control of data exchange in the fashion of a modem. Together these two levels represent a standard for future interface design specifications for the next generation of telecommunications interface cards and ports.

TABLE 9.1. Integrated Service Digital Network (ISDN) Standard

Location	*Designation*	*Task*
Hardware	Physical Layer	Direct electrical interface between computer and the local area network
Hardware	Data Link Layer	Data exchange in the fashion of a modem
Software	Network Layer	Assembles data packets in appropriate format
Software	Transport Layer	Divides and reassembles data packets
Software	Session Layer	Manages device addressing
Software	Presentation Layer	Translates data to and from the resident application format
Software	Application Layer	Accepts log-on requests and interprets the user's system requests, marshalling the system's resources to carry them out

At the software level are five layers of translation for data format compatibility. The *network layer* assembles packets in appropriate format. The *transport layer* controls the division and reassembly of data into packets. The *session layer* manages the placing of the packets on the network. The *presentation layer* accepts the user's incoming and outgoing packets and translates them to and from the resident application format. The *application layer* accepts log-on requests and interprets the user's system requests, marshalling the system's resources to carry them out. Together these five levels represent a standard for future communications software specifications that will be incorporated into the next generation of terminal programs.

Of the networking systems on the market today, most have begun to incorporate the ISDN, or its subsets, into their design. As a result, many

LAN systems can now be linked by way of *bridges* and *gateways*. A bridge is a link between similar LAN systems. The bridge manages the exchange of information between the systems by keeping track of the messages that are to be exchanged. A gateway acts much as a bridge, but also translates data to conform with speed and code requirements of other devices and systems.

An example of the development of networking is the Ethernet system. Designed by Xerox in 1976, Ethernet is a high-speed communications medium that permits data exchange at speeds up to 10 million bits per second. Adopted as a standard by the Institute of Electrical and Electronic Engineers (IEEE), almost every computer has access to Ethernet (Meng, 1988). This includes the supercomputers from Cray, the VAX® from Digital Equipment Corporation, the IBM PC, UNIX® workstations like those of Sun Microsystems and Apollo Computer, and AppleTalk® systems from Apple Computer.

Ethernet is configured as a bus: a single cable, open on both ends, to which network devices are attached. All that is necessary to connect a digital device to the Ethernet is a tap connector and an Ethernet transceiver. The Ethernet transceiver is an interface card implementing the ISDN standards to the extent necessary to be compatible with the digital device's native networking protocol. Examples of systems that can be networked via built-in Ethernet protocol bridges include the VAX DECnet®, Department of Defense TCP/IP (Transmission Control Protocol/Internet Protocol) systems, including UNIX systems and the IBM PC, and the XNS (Xerox Network Service) systems such as 3Com's 3Plus networking products. Apple Computer has recently introduced a line of Ethernet gateway interface products that permit AppleTalk networks to be compatible with Ethernet standards, including EtherTalk, EtherPort SE, EtherSC, and FastPath. It is rapidly becoming difficult to define the difference between LANs and conventional telecommunications as ISDN interface standards become incorporated into greater numbers of products, increasing the variety of interconnections possible among computer devices.

DISTANCE LEARNING: A BROADENING DEFINITION

The use of telecommunications to offer instruction to students removed geographically or "in class" at different times, or both, is known as distance education. As such, the subject has a history rich in innovative ideas and high hopes for educational change. While it is beyond the scope of this book to examine distance education in great depth, it is illustrative to consider some of the ways in which new technological developments are bringing renewed excitement and opening up new possibilities for educational innovation.

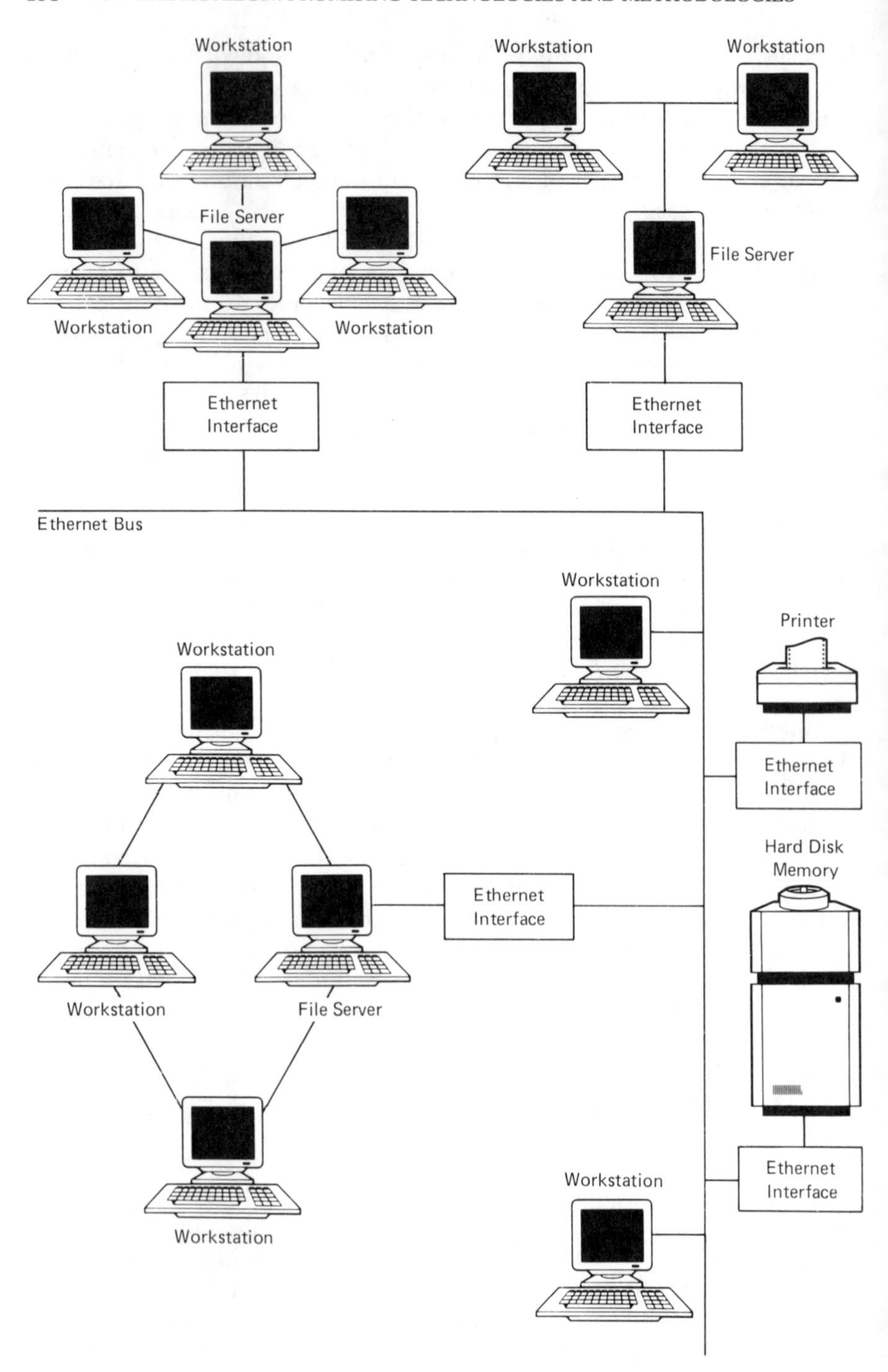

FIGURE 9.6. Ethernet

Educational Radio and Television

Educational radio and television have been around since the media were invented. Operating on the principle that pooled resources can bring high-quality educational programming into every classroom in the nation, the concept has stood the test of time. While never reaching the full potential for change in classroom practice envisioned by its proponents, it has none the less become a regular part of most school curricula. Approximately 70% of the nation's schools can receive broadcast instructional television, and studies by the U. S. Department of Education, National Center for Education Statistics (NCES), indicated that instructional television utilization averages 20 minutes per school day, or about 5% of available class time (Office of Technology Assessment, 1987).

Recently, Public Broadcasting Service Enterprises, a for-profit subsidiary of the Public Broadcasting Services launched in 1985 to generate revenue for PBS and public television, announced plans to introduce high-speed digital data delivery services using Vertical Blanking Interval (VBI) transmission on public television station signals (PBS Enterprises, 1987). Using a special receiver—the Vertical Blanking Interval Data Receiver (VDR)–designed in cooperation with EEG Enterprises, schools and homes will be able to decode data at 9600 baud. VBI is the time during which the electron scanning beam of a television cathode ray tube, which draws the television picture on the screen, is turned off to allow the beam to move back to the upper left-hand corner of the screen in preparation for scanning the next picture. Television pictures are generated by this scanning beam 30 times each second. If the beam is not turned off before moving back to the top, it leaves a streak of light across the screen. During this interval, no useful television information is being transmitted, and it is therefore available for other uses such as the VDR project.

VDR puts the PBS system in a unique position to offer low-cost digital data services to more than 80 million homes and schools through nationwide point-to-multipoint transmission of its signals. The potential for education is enormous. Combined with software that can be programmed to select packets on certain subjects, a computer could collect and compile the latest information on particular topics for students or class projects and have it ready and waiting until it is convenient for the users to transfer it to their own disks or systems.

Cable Communications Systems (CATV)

Cable Television (CATV) is another broadcast medium that is coming into its own. Responding to recent deregulation of the telecommunications industry, CATV companies can now compete for licenses to serve local communities. Many of the licenses include provisions for one or more public access channels and equipment and training for local residents in television production. The potential for educational programming for local residents is a largely untapped resource for meeting the needs of exceptional, homebound, and adult students.

Cable systems also are permitted to rebroadcast FM radio signals. The potential for the widespread introduction of digital data on subcarrier frequencies is enormous. When simulcast with educational programming, such digital transmissions offer the possibility for generating real-time "electronic chalkboards" on computer screens while a teacher presents a lesson on "live" television. In addition, the application of technology pioneered by the Qube system of Warner Communications of Columbus, Ohio, permits interactive communication between teacher and students on the digital computer channel.

Teleconferencing

Teleconferencing is a form of "meeting" among participants at separate locations and perhaps at different times. The concept represents a merging of the conference telephone call, electronic mail, and the computer bulletin board. It may not be long before it incorporates cable television capability as well. Participants typically log on from their home or office terminals on a regular basis to read and respond to contributions by others. The state of the art is such that effective and efficient information exchange, interaction, and decision making can be supported (Licklider, 1982).

The potential of teleconferencing lies in its ability to free meetings from two key constraints: the need for participants to gather at one place and time, and limitations on interaction that introduce inefficiency. The ability to connect computers by telephone permits participants to contribute from a distance and have their input stored for review by others. The telephone eliminates the problem of distance, and computer storage eliminates the problem of timing. The interactive advantages are not as obvious, but provide the potential for teleconferences to become superior to face-to-face meetings.

Three subtle constraints can be bypassed through teleconferencing. One is the limitation on the number of participants contributing at any one time. In a face-to-face meeting, only one person can speak at a time, while the computer can accept virtually simultaneous contributions from most any number of participants. A second is the limitation on information processing—speech cannot be processed by people as fast as printed information. Third, face-to-face conferences typically do not have good contact with the outside world, so the information that is available to the meeting is only that which is brought by the participants. With an online teleconference there is unlimited access to a wide variety of information, data, and processing resources. The critical question is whether these advantages are put to good enough use to make teleconferencing a viable meeting option.

Many educators are presently exploring this potential. FIDOnet, one of the largest grass-roots telecommunications networks in the country, provides nationwide teleconferencing to just about anyone who has a computer and modem. FIDOnet is a computer bulletin board program that has the ability to call other similarly equipped systems and forward messages. Each FIDOnet bulletin board is thus the center, or node, of what is effectively a star-type local area network. FIDOnet bulletin boards are located just about everywhere around the country, so a message posted in Boston may well be posted in Florida, California, and Kansas by the next day. Reply messages can be sent from any posting FIDOnet bulletin board.

The FIDOnet BBS software supports multiple bulletin boards that are selected by users from a main menu. These subboards are often known as forums, and are of two types: local and echo. A local subboard is just that—accessible only to the users calling the bulletin board directly. In general, this limits participation to those who live within the local calling area. An echo board, however, is a store-and-forward board—any messages posted on an echo board are passed along the FIDOnet system when the local BBS connects to another FIDOnet node. If the receiving node runs that same echo board, the messages are posted there. In any case, the messages are passed along the system. Generally, any message posted on an echo board is posted within two days on all other FIDOnet nodes around the country that run that same board.

FIDOnet's potential for providing a nationwide mutual support system for teachers of physics prompted Morton M. Sternheim, Professor of Physics at the University of Massachusetts, to establish The Physics Forum computer bulletin board (that runs OPUS, a "shareware" BBS compatible with FIDOnet). Initially designed to help eliminate the isolation of high school physics teachers (most schools have only one physics teacher, who often teaches physics as part of a larger science teaching assignment), the project was funded under a grant supporting efforts to improve science teaching. The concept caught on and, although the number of physics teachers turned out to be too small to sustain an active forum, other science teachers began to call the bulletin board, leading to the development of an active telecommunications conference on a number of topics. As a part of the FIDO network, The Physics Forum now supports echo boards for general science, college science teaching, education, astronomy, physics, biology; Apple, IBM, and Macintosh; as well as a number of boards of local interest. More than 1000 teachers have called the bulletin board at least once, with typically 200 different users a month and an average of 30 calls a day (Sternheim, 1988). See the Resource Section for more information on FIDOnet.

RECENT INITIATIVES

As technological capabilities expand, it is becoming more and more difficult to categorize the educational projects being designed to take advantage of the new systems. It is no longer possible to identify efforts as simply "educational television" or "teleconferencing." The ability to combine technologies has opened up whole new possibilities for education and given fresh life to older methodologies. Some sense of this exponential growth of educational technology can be gained from examining the concept of distance education from the perspective of the educational goal rather than the educational medium. Two longstanding goals, and one new one, are illustrative of the new life that telecommunications technology has brought to distance education.

Educational Equity for Rural Schools

Well established and funded due to their eligibility for state and federal support are efforts to use educational technology to provide educational equity for students in rural or isolated schools. The need to provide expert teachers to districts where few exist has provided the motivation for some very innovative combinations of telecommunications technology for the purpose of instruction. The longstanding problem with educational television and radio has been its non-interactive nature. Until very recently, telecommunications has meant telephone, an instructional method that lacks the critical visual component. Several recent projects have sought to eliminate that limitation.

The combination of video technology and the conference telephone call is being explored by the Stevens Institute of Technology in Hoboken, New Jersey. Funded by a grant from the New Jersey Department of Higher Education, the Stevens project uses prerecorded videotapes that can be controlled from the originating location. When "class" begins, the instructor can lecture, discuss and question students via the conference telephone connection, and control the operation of the video playback machines. This permits portions of the taped material to be reviewed as necessary. The students can ask questions and hold discussions with each other and the instructor. While this combination of technology does not permit the instructor and students to see each other or share drawings and notes, it does add structure and dialogue to the use of interactive television-type learning materials.

Projects, however, are under way in several states to combine the conference telephone call with the computer modem for interactive audiographic instruction. Students in Utah, New York, and Massachusetts are now able to study calculus in this fashion (Office of Technology Assessment, 1987). Known as *teleteaching,* instructors are able to talk with students and at the same time use computer graphics to illustrate concepts or write equations. Systems like the Harvard University Extension School teleteaching project (Hughes-Hallett, 1988), use a new type of modem produced by the Optel Corporation that can transmit both voice and computer data on one telephone line at the same time. Operating through a telephone company *bridge,* which looks like a telephone number to the students, the system becomes a local area network while class is in session. With the computer comes the ability to set up and display special screens, draw freehand with a light pen, save at any time, and clear and redisplay notes and drawings from any previously saved screen. Furthermore, both students and instructor can work with the computer display. The voice capability permits discussion and questions.

The limitation of teleteaching is that teacher and students cannot see each other. While the computer adds graphic capabilities—an important advancement over the conference telephone call alone—it works best with subject matter that lends itself to text or diagrams. To get around this problem, a number of projects are experimenting with satellite television links in combination with the conference telephone call. Oklahoma State University has begun to offer German language instruction to classes in 50 Oklahoma schools. Enrollment in these classes doubled in one year. In Utah, Nevada, Colorado, and Arkansas, a Spanish course is being taken by students in 26 schools. In both projects, the teacher is broadcast "live" over the satellite channel and the class converses over the telephone connection (Office of Technology Assessment, 1987). The video portion is only one way, but at least the students are able to see the teacher and his or her nonverbal information.

Students with Special Needs

Meeting the specialized needs of students is another established use of distance education. Whether this involves specialized courses for special education teachers, or the use of educational media to attract and hold the attention of children, the goal has been to provide the extra support that might not be available in every school. Telecommunications is creating new possibilities for working with exceptional children.

The Montgomery County Intermediate Unit has established a computer bulletin board to serve eleven of its special education classes in Norristown, Pennsylvania. Developed by teacher Fred Wheeler, the bulletin board supports the efforts of the school system to use writing as a means of drawing emotionally disturbed students into the mainstream of healthy social interaction. See Chapter 4 for more information.

A similar project, developed by the Eastern Connecticut (EAST-CONN) Regional Educational Service Center, has also used telecommunications as a link to the outside world for juveniles who are being detained. The opportunity for young people confined to locked residential settings to use telecommunications as a means of developing positive social skills has shown some promise. These children represent one of society's most perplexing group of students. They are often lacking in both self-esteem and literacy skills, and the challenge is to find ways to break the vicious circle that leads to further problems. When the problems lead to detention within a closed social system populated by others with similar deficiencies, it is difficult to provide an environment conducive to helping these youngsters. The anonymity of telecommunications permits these children to present a "blank slate" to their peers, and a chance to start healthy relationships without their background coloring the initial interactions. The possibilities for social rehabilitation hold exciting promise for this use of telecommunications.

Groupware

Beyond simple networking and the sharing of information and resources, however, lies the potential for important changes in the way people work and learn together. Groupware, or computer-supported cooperative work (CSCW), is software that assists working groups. Following the successful introduction of telecommunications or a LAN system, CSCW is emerging as a key application for addressing the needs and problems that arise for an interacting group of people. Once individuals discover the power and versatility of sharing information, the desire to use these same systems to facilitate collaboration naturally emerges. Groupware was originally developed in the 1960s in response to the need for collaboration among users of large mainframe computer systems and networks. After a decade of decline, it is now being rediscovered as a means of meeting similar needs among users of personal computer networks.

Over the last ten years, the personal computer has steadily brought more and more of the power of mini and mainframe systems to the individual user. Its standalone nature has provided accessibility to, and wider choice among, computing applications. What was originally sacrificed in terms of speed, power, and capacity has been rapidly recouped by technological developments. What was lost in terms of shared information and resources is only now beginning to be regained through telecommunications and LANs. E-mail has made possible a simple but effective form of information exchange among personal computer users. Groupware promises to provide sophisticated tools for efficient and effective collaboration among individuals using networked personal computers.

Some of the earliest work in the design of effective computer systems to support working groups was done in the mid-1960s by Douglas Engelbart at the Augmentation Research Center (ARC) at the Stanford Research

Institute International (Engelbart & Lehtman, 1988). The system that was developed was called NLS (On-Line System), and was designed to support collaboration among people doing their work in an asynchronous, geographically distributed manner. The system pioneered many of the techniques that are in widespread use in personal computers today, including windowed screen display, mouse-controlled cursors, hypertextual document linking, and standard human interface guidelines for all applications. The importance of the work, however, lay in more than its technical developments. The goal was the augmentation of knowledge workers with technological tools. Toward that goal, NLS identified a number of important issues that surround effective computer-supported collaborative work, including support for collaborative dialogue, document development, research intelligence, community handbook, computer-based instruction, meetings and conferences, community management and organization, and special tools. These elements of collaborative work are being rediscovered and utilized in the new generation of groupware products for networked personal computers.

Computer support for collaborative dialogue included tools for message composition, review, cross-referencing, modification, transmission, storage, indexing, and retrieving. NLS also automatically sent and received messages, notified recipients, stored and indexed the contents, and kept a running history of exchanges. The goal was to provide a service a step beyond E-mail in sophistication that effectively managed the dialogue among a community of workers sharing the same knowledge base.

Document development support addressed the need for managing the control of composing, studying, and revising documents from their working stage through final copy. Research at ARC found that when many authors and editors became involved in the production of a document, a means of organizing the process was essential to effective project completion. Among the needs addressed were the ability to maintain a complete history of drafts, establish write-protection for the current working version, and provide the ability to merge, annotate, and update revised copies. The goal was to enable a group of people to review, comment, and revise a document entirely online, thereby avoiding the problems and inefficiencies inherent in having to print, circulate, track, collect, collate, and retype revisions.

Research intelligence was provided so that users could actively monitor the data stream for references and citations being developed and recorded elsewhere in the system that might be useful to their own particular work. These tools permitted users to automatically build their own research catalogs to both internal and external information. The result was the creation of a dynamic and highly useful intelligent database that kept users up to date on external activities that might affect their activities.

The community handbook was a service to coordinate the large and complex body of documentation and associated references that a working group develops in the course of its activities. The system was designed to automatically compile and index the group's principles, practices, standards, goals, techniques, methods, and positions. The result was a resource supporting an active dialogue about the nature of the working group and its community positions that was separate from the actual work of the group, but essential to its effective functioning.

Computer-based instruction addressed the reality of different levels of expertise within the working group. The system provided help and tutoring, as well as the ability to use shared screens to "follow along with an expert," to permit individual users to develop skills as it became necessary. This function also included dynamic access to the community handbook so that skills could be understood in the context of the community's behaviors and beliefs as a whole.

The meeting and conference feature permitted both local and distributed sessions. Video overlay and split-screen displays were pioneered as a means to facilitate group work with common data. The feature proved useful for facilitating the dynamic control of agenda and the collaborative creation of position papers.

Community management and organization features augmented traditional computer supported planning and management operations. The system provided the ability to automatically build histories of dialogues concerning plans, commitments, schedules, and specifications for the participants. The goal was to provide better and more explicit tracking of action oriented conversations within an organization.

Finally, subgroups within the community often have need for special tools or applications to facilitate their work. NLS was designed to provide these special services. The ability of each user to customize the system to his or her own particular needs was a major contribution to the success of the overall program. In effect, the user had the ability to individualize the system and at the same time maintain access to all the shared resources and contacts that the system made possible. It is not surprising, therefore, to see these same principles for the effective coordination of shared work reemerging in the groupware applications that are being developed for networked groups of personal computer users. The personal computer provides the advantages of individual customization of applications, while groupware supplies the tools and supporting environment for access to shared resources and collaborative work.

In general, most groupware products for networked personal computers have focused on document production and the management and organization needs of collaborating groups. At present, no products address all the needs that the research at ARC articulated during the development of NLS, or provide the comprehensive services of that system to networked personal computer users. Of the groupware products that have been introduced over the past several years, several have achieved some success in the marketplace. Among these are ForComment from Broderbund, Higgins by Conetic Systems, The Coordinator by Action Technologies, Caucus by Metasystems Design Group, Office Works by Data Access, and WordPerfect Office by WordPerfect Corporation (Opper, 1988, p. 275).

ForComment is a document production product that runs on IBM PCs or compatibles and can support up to 16 contributors. The program allows multiple reviewers to comment and see each others' remarks without actually altering the original document. The ability to change the document rests with the designated author, who can automatically merge all the comments into a new working copy and decide which to include or exclude, entirely online. Commenter's suggestions are initialed, and a report function can track all review activity.

The other products are management and coordination applications for working groups. Higgins is designed around relational databases, The Coordinator makes the intent of action-oriented communications explicit to users, Caucus supports extensive conferencing activity, Office Works makes extensive organizational use of the telephone message, and WordPerfect Office centers on the calendar/scheduler function. All run on IBM PC LAN systems, all provide access to shared databases, and all track and keep records of message exchanges.

So far, groupware products are special applications that enhance certain aspects of group collaboration within particular working environments. The power of any of them is best tapped when the group's functioning style is much like that for which the product is intended. To date, no generic, general purpose groupware tool like NLS has appeared, but it is not hard to imagine that such a product will be developed for personal computer networks in the near future.

ForComment offers the most obvious educational applicability at present. The ability to collaborate online on the production of written work has great potential for school writing laboratories. Both students and teachers could share their work for group comment and interaction. Real-time online comments and editing promises to bring the power of the computer to the use of writing samples in classrooms. The integration of groupware concepts into the school setting can reasonably be expected to follow the present trend of creating LAN systems.

We can only imagine how teaching and learning will evolve once full-scale NLS type groupware, with its ability to support collaborative activity in both real-time and asynchronous styles, either locally or geographically distributed, becomes available. The potential for "schools without walls" and individualization without losing the critical social context factors presently inherent in networked learning is on the horizon. Groupware is the next step in telecommunications and networking. Now that individuals can use telecommunications to communicate while distributed over both distance and time, it is time to develop the means to coordinate collaborative learning environments utilizing these systems.

COMPUTING IN THE YEAR 2000

Apple Computer Company's Project 2000, begun in 1987, is a competition among universities to come up with a vision of what computing will be like in the year 2000. The University of Illinois is the most recent winner of the competition, with their proposal for TABLET, a computer with the following features:

- The size of a standard notebook (approximately 8 × 11 inches) and weighing only a few pounds.
- A touch-sensitive screen that will function as the primary input/output device.
- System support for video, image, graphics, data, and text data that will all be available in color.
- Handwriting recognition software that will allow the use of a pencil-like stylus as an input device.
- Voice recognition and speech-to-text software that will allow the use of the system for dictation without the need for transcription.
- Data storage that will take the form of "laser cards," erasable optical RAM. The size of a credit card, each laser card will hold a gigabyte of information, and will have no moving parts to wear out.
- Communications that will be provided through a built-in cellular radio that allows the transmission of both voice and data.
- Communications with peripheral devices such as printers, projectors, and video cameras that will be provided through an infrared interface similar to today's television remote control devices. (Drazen, 1988, p. 3)

Sound farfetched? Maybe, but realize that all of the technology that makes up TABLET is available in some form today.

CREATING GLOBAL CLASSROOMS

A long held dream of many educators is the realization of a "global community." Modern telecommunications is beginning to offer practical opportunities for teaching and learning without regard to political or geographic boundaries. One such experimental project was developed by University Tech-Tel (UTT) in association with Optel Communications Inc., Harvard University, Boston University, Beijing Normal University, the U. S. State Department, and its Chinese counterpart (University Tech-Tel Corp., 1988). The purpose of the project was to determine whether it is possible for Chinese students to learn about the latest developments in their fields from recognized American experts without either group having to leave home. Utilizing the same audio-graphic "teleteaching" methodology developed in Cambridge to offer the Harvard Extension School Calculus course, two five-week courses were offered, one in Molecular Biology by Boston University professors, and the other in Artificial Intelligence Expert Systems by Harvard University professors. Participants felt the program was highly successful, especially in its ability to provide high-quality interaction and the sharing of text and diagrams, all without the time, expense, and logistical problems associated with distant travel. Plans are under way for routine educational exchanges of this sort between the two countries.

The potentialities, however, are proving to be multidisciplinary. A highly successful interactive satellite course on the history of the nuclear arms race was jointly conducted by Tufts University of Medford, Massachusetts, and the State University of Moscow. Conceived by Professor Martin Sherwin, Director of the Tufts Nuclear Age and Humanities Center, the project provided students in both countries the opportunity to hear and discuss simultaneous interpretations of three aspects of the nuclear age with present-day policy makers and original participants in the development of atomic capabilities. Conducted in three Saturday sessions, one focused on the early history of atomic research in each country, another on the development of nuclear strategic thinking, and the third on the Cuban missile crisis. As an illustration of the potential of telecommunications to bring people together, not all the panelists were participating from the seminar sites. Several panelists participated in the discussion via remote television hookups (Sherwin, 1988).

An even more ambitious project is the Global University. Begun in 1972 as the Global Systems Analysis and Simulation Project for global peace gaming, the idea has evolved to creating a full Global (electronic) University.

> The GU is directing itself to four essential goals: The globalization of educational opportunities for all; support of research and development; the use of global-scale tools such as peace-gaming and global village meetings so as to explore new alternatives for a world whose people are interdependent; and the globalization of employment opportunities.
>
> [This organization] is incorporated as a New York, non-profit, educational service organization to assist and enhance the quality and availability of international educational exchange. (Utsumi, *et al.*, 1989, p. 75)

The possibilities for improving education and instructional practice through telecommunications are unlimited. By offering quality instruction to rural schools; providing new ways to help disturbed children; helping break down the barriers of time, space, culture, and ideology; and finding more effective ways for groups to work together, developments in telecommunications technology are being harnessed for educational purposes. Whether it's a teacher using a modem purchased at a local computer store, or the director of a multimillion dollar government project using the latest experimental technology, educators are developing new and better ways to bring high-quality learning environments to all students. The future holds great promise, and there is every reason to be optimistic about the contribution of telecommunications toward achieving the goal of global education and understanding.

REFERENCES

Charp, S., & Hines, I. J. (1988, April). The basic principles of networking. *T.H.E. Journal*, pp. 94–98.

Computer data and pictures can now be transmitted via radio. (1984, December). *Radio Only*, p. 15.

Drazen, A. (1988, July) *I/O*, p. 3.

Earth Lab integrated software. (1987, Summer). *Teaching, Learning, & Technology*, pp. 27–29.

Englebart, D., & Lehtman, H. (1988, December). Working together. *Byte*, pp. 245–252.

Hughes-Hallett, D., Mathematics Dept., Harvard University. (1988, July). Personal correspondence.

Johnson, N. (1988, March). Local area networks and language learning. *The Computing Teacher*, *15*(6), 8–10.

Joyce, E. (1985, May 28). Software takes to the air. *PC Magazine*.

Licklider, J. C. R. (1982). *Teleconferencing, innovations in telecommunications*, pp. 949–993. Orlando, FL: Academic Press.

Meng, B. (1988, January). The Ethernet solution. *Macworld*, pp. 128–137.

New England Telephone. Press release. (date not available)

Newman, D. (1986, December 31). *Earth Lab: Progress report*. New York: Bank Street College of Education.

Office of Technology Assessment, U. S. Congress. (1987, March). *Trends and status of computers in schools*. Staff Paper, pp. 120–129. Washington, DC: Author.

Opper, S. (1988, December). A groupware toolbox. *Byte*, pp. 275–282.

Packet, J., Eastham Elementary School, Eastham, MA. (1988, July). Personal correspondence.

Public Broadcasting Service Enterprises. (1987, March 13). Press release.

Seiger, B. (1988, Spring). Radio telecommunications. *Hands On*, pp. 13 & 22.

Sherwin, M., Director of Tufts Nuclear Age and Humanities Center, Tufts University, Medford MA 02155. (1988, July). Personal correspondence.

Smirnoff, S. R. (1987, June). Linking the last frontier with the land down under. *Alascom Spectrum*, pp. 9–11.

Sofcast Corporation, 2565 152nd Ave., N.E., Redmond, WA 98052. Press release (undated).

Sternheim, M. M., University of Massachusetts, Amherst, MA. (1988, July). Personal correspondence.

Stoner, D. (1986, January). *Personal Communications Technology*, pp. 43–44.

University Tech-Tel Corporation. (1988, March 15). Press release.

Utsumi, T., Rossman, P., & Rosen, S. (1989). Global education for the 21st century: The GU consortium. *T.H.E. Journal, 16*(7), 75–77.

Resource Section

The information in this section is provided to aid the reader in accessing the products and services mentioned in the book, as well as other popular or useful items in each category. The section is divided into seven areas:

Associations, Universities, and Organizations
Bulletin Boards (both public and private)
Commercial Networks and Databases
Hardware
Publications
Software
Teaching Materials

The information is as current as the production schedule of a book allows. Each citation includes the name, address, a brief description, and the chapter in which the item is mentioned if included in the text. If something mentioned in the text is omitted from this section, it is because information about the item was not available at press time or the project is no longer active.

ASSOCIATIONS, UNIVERSITIES, AND ORGANIZATIONS

American Radio Relay League
ARRL
225 Main Street
Newington, CT 06106
203-666-1541
A non-commerical association of radio amateurs organized for the promotion of amateur radio communications, experimentation, and the establishment of networks so as to provide communication in the event of disasters or other emergencies. (Chapter 9)

Association for Educational Communications & Technology (AECT)
1126 16th Street, NW
Washington, DC 20036
202-466-4780
AECT's commitment is to improve education through the carefully designed application of technologies to the teaching/learning process. It supports an educational network that includes bulletin boards and electronic mail. Its publications are *Educational Technology Research & Development* and *Tech Trends.*

Bank Street College of Education
610 West 112th Street
New York, NY 10025
212-663-7200
The Center for Children and Technology is a research institute exploring the contributions of new technologies o learning. This organization produced the EarthLab, a local area network for augmenting the study of science. Director: Dr. Karen Sheingold

IEEE (The Institute of Electrical and Electronic Engineers, Inc.)
345 East 47 Street
New York, NY 10017
212-705-7900
One of the world's largest professional organizations, primarily servicing the needs of electrical engineers and related professionals. (Chapter 9)

International Council for Computers in Education
University of Oregon
1787 Agate Street
Eugene, OR 97403
503-686-4414
Devoted to the interests of teachers and technology, this organization is composed of representatives from Computer Using Educator (CUE) groups around the world. ICCE sponsors several special interest groups and a host of publications. It is the co-sponsor of the National Educational Computing Conference. Its official journal is *The Computing Teacher*. As of June, 1989, ICCE and IACE merged to become the International Society for Technology in Education (ISTE).

Lesley College
29 Everett Street
Cambridge, MA 02138-2790
1-800-999-1959
Offers a Master's Degree in Computers in Education which includes several courses dealing with telecommunications and other technologies applicable to education. Director: Nancy Roberts

Massachusetts Corporation for Educational Telecommunications
World Trade Center, Suite 115
Boston, MA 02210
617-439-5888
In collaboration with the other New England states and New York, this organization is developing a long-term comprehensive plan for a regional educational telecommunications net-

work. Executive Director: Richard A. Borten

National Council for the Social Studies
3501 Newark Street, NW
Washington, DC 20016
202-966-7840
A nationwide organization focusing on the needs of social studies teachers and teacher educators. The official journals of this organization are *Social Education,* published seven times a year, and *Social Studies and the Young Learner,* published four times a year.

National Council for the Teachers of English
1111 Kenyon Road
Urbana, IL 61801
217-328-3870
An association that serves the individual teacher of English as well as the collective goals of the English teaching profession. Its major publications are *Language Arts* for elementary teachers, *English Journal* for middle school through high school teachers, and *College English* for the college scholar-teacher. Each journal is published nine times a year. (Chapter 5)

National Council for the Teachers of Mathematics
1906 Association Drive
Reston, VA 22091
703-620-9840
MIX:nctm
CompuServe 75445,1161
The major national organization for teachers of mathematics and educators of mathematics teachers, its official journals are: *Journal for Research in Mathematics Education,* published five times per year; *Mathematics Teacher,* focusing on secondary mathematics, published nine times per year; and *Arithmetic Teacher* for elementary teachers, published nine times per year.

National Science Teachers Association
1742 Connecticut Avenue, NW
Washington, DC 20009
202-328-5800
Publishes *Science and Children* for elementary teachers, *Science Scope* for middle school teachers, *The Science Teacher* for high school teachers, and *The Journal of College Science Teaching,* each nine times a year. This organization attempts to meet the needs of the whole range of people involved in the teaching of science.

NYNEX
Public Service Center
6 St. James Avenue
Boston, MA 02116
617-956-8400
The regional telephone operating company for the New York and New England area. (Chapter 9)

Oklahoma State University
Department of Curriculum and Instruction
College of Education
Stillwater, OK 74078-0146
405-642-7125
An innovator in applying technology, especially teleconferencing, to education.
Contact: Joyce Friske

San Diego State University
Educational Technology Department
San Diego, CA 92182-0311
619-594-6718
Supports many telecommunications in education projects, especially on the West Coast.
Contacts: Bernie and June Dodge

Stevens Institute of Technology
Castle Point
Hoboken, NJ 07030
201-420-5239
One of the first universities to require their students to purchase computers,

a major innovator in educational technology.

Technical Education Research Centers (TERC)
1696 Massachusetts Avenue
Cambridge, MA 02138
617-549-3890
A non-profit educational research organization which sponsors Kids Network, TERC's Star Schools project, and LABNET. Director: Robert Tinker

University Tech-Tel Corporation (UTT)
4720 Montgomery Lane, Suite 1100
Bethesda, MD 20814
301-652-0871
Founded in 1985, UTT installs and maintains educational and technological transfer systems for universities and research institutions whose telecommunications needs require personal interaction, both voice and data. (Chapter 9)

BULLETIN BOARDS

BreadNET
Bread Loaf School of English
Middlebury College
Middlebury, VT 05753
802-388-3711
A network of teachers who have taken part in the Bread Loaf summer teaching of writing program at Middlebury College. (Chapter 5)

Champlain Valley Union High School Electronic Bulletin Board
Craig Lyndes
Champlain Valley Union High School
RR2, Box 160
Hinesburg, VT 05461
802-482-2101
An example of a student-run bulletin board. (Chapter 3)

Computer Pals
See *Online* under ***Commercial Networks.*** (Chapter 5)

FIDOnet
Bulletin board system that has the ability to link together similarly equipped bulletin boards, thus creating an international network, currently composed of 4500 boards. Access to FIDOnet is only through local participating bulletin boards. Operates through a list of nodes that is updated weekly by a hierarchical structure of volunteers. (Chapters 7 and 9)

FrEdMail (Free Educational Electronic Mail Network)
Al Rogers
4021 Allen School Road
Bonita, CA 92002
Modem 619-292-1816
CompuServe 76167, 3514
A network of school-based bulletin boards that allows schools to communicate with each other through node-designated bulletin boards that transfer files across the country through the node system. (Chapters 3 and 5)

Imagination Network
Fred D'Ignazio
Multi-Media Classrooms, Inc.
1302 Beech Street
East Lansing, MI 48823
517-337-1549
A network that joins together the schools participating in the Multi-Media Classroom project. (Chapter 5)

InterCultural Learning Network
The classes involved in this network have now joined either the FrEdMail Network or the AT&T Long-Distance Learning Network. (Chapters 5 and 6)

LES-COM-net
George Willet
3485 Miller Street
Wheatridge, CO 80033
Modem 303-233-5824
A regional bulletin board for teachers and teacher/students in Lesley College Graduate Programs. (Chapters 2 and 3)

MeLink
Cathy Glaude
Maine Computer Consortium
P.O. Box 620
Auburn, ME 04210
207-783-0833
Modem 207-783-9776
An example of a consortium bulletin board serving the students and teachers in Maine. (Chapters 3, 4, and 5)

SeniorNet
Lone Mountain Campus
Rossi Wing
University of San Francisco
San Francisco, CA 94117-1080
415-666-6505
Designed especially for people over 60, the network is part of the Computers for Kids Over Sixty Project in San Francisco. (Chapter 3)

COMMERCIAL NETWORKS AND DATABASES

Accu-Weather
619 College Avenue
State College, PA 16801
814-237-0309
A national online database that also has available a software package, *Accu-Weather Forecaster,* that is designed to be used with the database. (Chapter 6)

AlaskaNet
See *OnLine.* (Chapter 5)

Alascom
See *OnLine.* (Chapters 5 and 9)

AppleLink-Personal Edition
8619 Westwood Center Drive
Vienna, VA 22180
800-227-6364
A telecommunications network designed for educators. (Chapters 4 and 5)

ARPANET (Advanced Research Projects Agency Network)
Stanford Research Institute
Menlo Park, CA 94025
800-235-3155
The first computer network, still used primarily for communications within the scientific community. (Chapters 1 and 6)

AT&T Long-Distance Learning Network
Cynthia Brinkman
AT&T
295 North Maple Avenue, Room 6234S3
Basking Ridge, NJ 07201
201-221-8544
The Long-Distance Learning Network is an innovative educational application of AT&T electronic mail designed for grades 3–12. Communications software and curriculum guides enable teachers and students to work with distance peers toward shared educational goals. (Chapter 6)

BITNET
EDUCOM
P. O. Box 364
Princeton, NJ 08540
603-520-3377
The international network of computers that links higher education institutions and certain other educa-

tional and research organizations. (Chapters 1 and 6)

Boston CitiNet
World Trade Center, Suite 717
Boston, MA 02210
617-439-5678
Modem 617-439-5699
A regional bulletin board that offers educators special rates. (Chapters 3 and 5)

Bibliographical Retrieval Service (BRS)
1200 Route 7
Latham, NY 12110
800-468-0908
Provides access to over 100 different electronic databases. (Chapter 4)

Carrier Services
Telenet 703-689-6300
Tymnet 800-368-3180
Uninet 800-821-5340
These three services allow peopie to access most of the other services listed here through a local, rather than a long-distance, telephone call. Generally, each of the other services give the needed access information in their materials. (Chapters 1 and 4)

CompuServe
5000 Arlington Center Blvd.
Columbus, OH 43220
800-848-8990 Customer Service
614-457-8600 in Ohio
One of the larger commercial services offering a wide variety of services and topics. Often sponsors educationally oriented forums. Hosts an Education Forum and a Foreign Language Educator's Forum. (Chapters 4, 5, and 6)

Delphi
General Videotex Corporation
3 Blackstone Street
Cambridge, MA 02139
617-491-3393
800-544-4005
A full-service network offering access to databases, electronic mail, and a variety of forums. (Chapter 6)

Dialog Information Services, Inc.
3460 Hillview Avenue
Palo Alto, CA 94304
415-858-2700
800-3DIALOG
Hundreds of databases are available through DIALOG. Also available is access to smaller numbers of databases at lower hourly costs designed for school use through DIALOG's Classmate. Classmate is part of DIALOG's Classroom Instruction Program (CIP) that also provides curriculum materials for introducing online searching to students. (Chapters 4 and 6)

Dow Jones News/Retrieval Service
P.O. Box 300
Princeton, NJ 08543-0300
800-522-3567 Membership Information
609-520-8349 in New York
The electronic service for the business community, it also offers educational ideas for all grade levels included in its *Educator's Guide*. Schools can negotiate flat yearly fees. (Chapters 4 and 6)

ERIC (Educational Resource Information Center)
See *DIALOG* or *ERIC MICROsearch* under ***Software.*** (Chapters 4 and 7)

GEnie (The General Electric Information Service)
410 North Washington Street
Rockville, MD 20850
800-638-9636
800-638-8369
A multipurpose electronic service. (Chapter 4)

LEXIS
See *NEXIS*. (Chapter 4)

NASA Spacelink
Modem 202-895-0028

Established for teachers and researchers, provides current and historic material on NASA aeronautics and space research activities. Also provides suggested classroom activities to involve students in science. (Chapter 6)

National Geographic Kids Network
National Geographic Society
Educational Media
17th and M Streets, NW
Washington DC 20036
800-638-4077
Middle school science materials focusing on collaborative scientific data collection by classrooms around the world. (Chapter 6)

National Geographic Weather Machine
See *National Geographic Kids Network*. A database providing a variety of weather information. (Chapter 6)

NewsNet
945 Haverford Rd.
Bryn Mawr, PA 19010
800-345-1301
800-537-0808 (Pennsylvania)
A news service for the business and professional communities. (Chapter 4)

NEXIS
Mead Data Central
9443 Springboro Pike
P. O. Box 933
Dayton, OH 45401
800-227-4908
A full-text news service that also offers specialized services for the law community (LEXIS) and for the medical community (MEDIS). (Chapter 4)

NSFNET
National Science Foundation
1800 G Street, NW
Washington, D. C. 20550
202-357-9717
A network established to link together the supercomputer centers placed by NSF around the country with students and scientists wishing to access them. (Chapters 1 and 6)

OCLC (Online Computer Library Center)
6565 Frantz Road
Dublin, OH 43017-0702
800-848-8286 (OH)
800-848-5878
A non-profit membership organization that provides automated services to libraries. (Chapter 1)

OnLine
Alascom, INC.
210 E. Bluff Road
P. O. Box 196607
Anchorage, Alaska 99519-6607
800-478-6500 (Alaska)
800-544-2233
An electronic mail service that allows Alaskans to connect to the world through AlaskaNet, a public packet network. (Chapters 5 and 9)

PCPursuit
12490 Sunset Valley Road
Reston, VA 22096
800-TELENET (800-835-3638)
Modem 800-835-3001
A service of Telenet, allowing unlimited access to a limited number of cities at a fixed monthly charge. (Chapter 4)

Prodigy Services Company
445 Hamilton Avenue
White Plains, NY 10601
914-993-8000
A joint venture of IBM and Sears, Prodigy's goal is to serve 10 million users. Because of some of its features, such as its flat monthly charge and graphics interface, it has potential for the education market. (Chapter 4)

RLG (Research Libraries Group)
Jordon Quadrangle
Stanford, CA 94305
415-962-9951

A non-profit organization of major U. S. universities and research institutions. RLG's automated information system, RLIN, combines databases and computer systems to support the partnership's cooperative programs. RLIN is a nationwide network, serving the materials processing and public services requirements of RLG's members and many non-member institutions and individuals. (Chapter 1)

RLIN (Research Libraries Information Network)
See *RLG*. (Chapter 1)

SpecialNet
Buck Schrotberger
Colorado Department of Education
201 E. Colfax
Denver, CO 80203
303-866-6722
A network devoted to the needs of special education teachers and administrators. (Chapter 7)

The Source
1616 Anderson Road
McLean, VA 22102
800-336-3366 Membership Information
800-336-3330 Customer Support
One of the larger, multipurpose electronic services with a section for teachers, students, and people concerned with special education. (Chapters 4 and 5)

VU/TEXT Information Service
325 Chestnut Street, Suite 1300
Philadelphia, PA 19106
800-323-2940
A service offering access to news articles from magazines, newswires, and newspapers. (Chapter 4)

WLN (Western Library Network)
Washington State Library
AJ-11
Olympia, WA 98504
206-459-6518
A subagency of the Washington State Library that provides computer services to libraries, primarily in the northeast U.S. (Chapter 1)

HARDWARE

Apple Computer, Inc.
20525 Mariani Avenue
Cupertino, CA 94014
800-538-9696
Their Personal Modem is a 300/1200 baud modem sold for around $400. (Chapter 2)

Corvus Systems Inc.
160 Great Oaks Blvd.
San Jose, CA 95119-1347
408-281-4100
Markets the Omninet local area network, often found in educational settings, as well as Constellation III software. (Chapter 9)

Ethernet
Xerox Corporation
1800 Longridge Road
P.O. Box 1600
Stamford, CT 06904-1600
800-334-6200
A networking system that provides the backbone, both hardware and software, for devices to communicate electronically. Adopted as a standard by the IEEE. (Chapter 9)

Fox Research 10Net
10 Net Communications
7016 Corporate Way
Dayton, OH 45459
800-358-1010

Provides networking products for IBM machines. (Chapter 9)

Hayes Microcomputer Products
705 Westech Drive
Norcross, CA 30092
404-449-8791
Offers seven different models ranging in price from $200 to $800 for a wide range of computers and ports. Creator of the Hayes Smartmodem Command Set. (Chapter 7)

IBM Corporation
Old Orchard Road
Armonk, NY 10504
914-765-1900
Manufacturers of the IBM PC family of computers, among a wide variety of other products. (Chapter 2)

Novell, Inc.
122 E. 1700 South
Provo, UT 84601
801-922-0354
A local area network used with IBM and IBM compatible machines. (Chapter 9)

Optel Communications
322 8th Avenue, Room 1400
New York, New York 10001
212-741-9000
Produces a modem that can transmit both voice and computer data on one telephone line at the same time. (Chapter 9)

Super Serial Interface Card
20525 Mariani Avenue
Cupertino, CA 94014
800-538-9696
The peripheral device needed to connect the Apple computer to the Apple Personal Modem. (Chapter 2)

Tandy Corporation
1800 One Tandy Center
Fort Worth, TX 76102
817-390-3700
Manufactures the Tandy computers and peripherals, all IBM compatible. (Chapter 5)

Warner Cable Communications
930 Kinnear Road
Columbus, OH 43212
614-481-5000
Manufacturer of the Qube System. (Chapter 9)

PUBLICATIONS

Apple Education News
Apple Computer, Inc.
20525 Mariani Avenue
M/S 23DE
Cupertino, CA 95041
A topically oriented newsletter focusing on educational computing published occasionally.

Classroom Computer Learning
Peter Li Inc.
19 Davis Drive
Belmont, CA 94002
415-592-7810
Attempts to link computer-based instruction with traditional classroom teaching; published eight times a year.

DBKids
Beverly Hunter
Targeted Learning Corporation
Route 1, Box 190
Amissville, VA 22002
703-937-4744
A newsletter reporting on database and networking activities for learning and teaching.

Electronic Learning
Scholastic, Inc.
730 Broadway
New York, NY 10003
212-505-3000
This journal provides non-technical introductions to educational computer applications.

FrEdMail News
See *FrEdMail* under ***Bulletin Boards.*** A newsletter for the FrEdMail Network, reporting on the services of the network and activities of the classes participating in it; published 4 times a year.

Online Today
CompuServe Inc.
5000 Arlington Centre Blvd.
Columbus, OH 43220
614-457-8600
The journal for CompuServe. Often has articles of interest to educators.

Teaching and Computers
Scholastic, Inc.
730 Broadway
New York, NY 10003
212-505-3000
Provides information and practical suggestions for integrating computers into the elementary classroom.

T.H.E. Journal
Information Synergy, Inc.
2626 S. Pullman
Santa Ana, CA 92705-0126
714-261-0366
Focusing on all levels of education, this journal contains a wealth of information on products and projects in each of its 10 issues per year.

The Computing Teacher
University of Oregon
1787 Agate Street
Eugene, OR 97403
503-686-4414
A general magazine for any teachers using technology in an educational setting; published nine times a year.

SOFTWARE

Apple Access II
Apple Computer, Inc.
20525 Mariani Avenue
Cupertino, CA 95041
800-538-9696
Telecommunications software designed specifically for the Apple family of products. (Chapter 2)

AppleWorks®
CLARIS
440 Clyde Avenue
Mountain View, CA 94043
415-960-1500
An integrated word processor, database, and spreadsheet software package for the Apple II family of computers. (Chapters 2 and 5)

Computer Chronicles Newswire
InterLearn, Inc.
Box 342
Cardiff by the Sea, CA 92007
619-942-0734
Provides writing and telecommunications software that allows classes to participate in the network either through The Source or through the mail. (Chapter 5)

Constellation III
See *Corvius* under ***Hardware.*** Network software for the Corvus System. (Chapter 9)

CMS SchoolNet
CUE Softswap
P.O. Box 271704
Concord, CA 94527-1704
415-685-7265
Particularly active on the west coast, CMS is a bulletin board software program that many telecommunicating teachers are using across the country. Mail is forwarded from area computers to designated nodes. From each node it is forwarded to the next correct node and then to the local bulletin board of the addressee. See *FrEdMail* under ***Bulletin Boards.*** (Chapter 5)

Einstein
Learning Link
356 West 58th Street
New York, NY 10019
212-560-6674
Provides access to many databases through one standard interface. Learning Link is owned by the public television station WNET 13. (Chapters 4 and 7)

Electronic Mailbag: An Electronic Mail Simulator
Exsym
301 North Harrison Street
Building B, Suite 435
Princeton, NJ 08540
609-737-2312
Runs on a single computer and allows up to 100 students to participate in electronic mail. (Chapter 3)

Electronic Village
Exsym
301 North Harrison Street
Building B, Suite 435
Princeton, NJ 08540
609-737-2312
A simulator of a 300-baud bulletin board system. (Chapter 3)

ERIC MICROsearch
Information Resources Publications
036 Huntington Hall
Syracuse University, NY 13244-2340
315-443-3640
A simulator for the ERIC bibliographic database. (Chapter 3)

FrEdWare
Includes FrEdWriter, FrEdSender, and FrEdFiler. See FrEdMail under ***Bulletin Boards.*** (Chapters 3 and 5)

Grolier's Electronic Encyclopedia
Grolier Electronic Publishing, Inc.
95 Madison Avenue
New York, NY 10016
212-696-9750
A CD-ROM containing the complete text from the *Academic American Encyclopedia*. Used in conjunction with a computer and a CD-ROM drive, information can be searched, saved, and then used with a word processor. (Chapter 4)

GroupWare Products
The following vendors, all mentioned in Chapter 9, provide the indicated groupware-type product:

Caucus
Metasystems Design Group Inc.
2000 North 15th Street, Suite 103
Arlington, VA 22201
703-739-5000

ForComment
Broderbund Software Inc.
17 Paul Drive
San Rafael, CA 94903
415-492-3200

Higgins
Conetic Systems Inc.
1470 Doolittle Drive
San Leandro, CA 94577
415-430-8875

Office Works
Data Access Corporation
14000 Southwest 119th Avenue
Miami, FL 33186
305-238-0012

The Coordinator
Action Technologies
2200 Powell Street, Suite 1100
Emeryville, CA 94608
415-654-4444

WordPerfect Office
WordPerfect Corporation
1555 North Technology Way
Orem, UT 84057

Information Connection
Grolier Electronic Publishing
Sherman Turnpike
Danbury, CT 06816
800-858-8858
A database search simulator. (Chapter 3)

KidMail
CUE Softswap
P.O. Box 271704
Concord, CA 94527-1704
415-685-7265
An electronic mail simulator. (Chapter 3). To participate in the KidMail class exchange, contact:
Wayne Ayers
Culver City High School
4401 Elenda St.
Culver City, CA 90230

Modemless CMS
CUE Softswap
P.O. Box 271704
Concord, CA 94527-1704
415-685-7265
A simulator useful to people who intend to use CMS.

OPUS
Boston Computer Society (BCS), Telecommunications SIG
One Center Plaza
Boston, MA 02108
617-367-8080
A shareware telecommunications package available from the BCS and other organizations that distribute such software. (Chapter 9)

Procom Plus
Datastorm Technologies, Inc.
P. O. Box 1417
Columbia, MO 65205
314-449-7012
A telecommunications package for the IBM PC and compatibles. (Chapter 2)

Real Time Writer
Real Time Learning Systems
2700 Connecticut Avenue
Washington, DC 20008-5330
202-483-1510
A network software that currently runs on the Fox Research Net10, Novell, and IBM token ring configuration. (Chapter 9)

Red Ryder
The FreeSoft Company
10828 Lacklink
St. Louis, MO 63114
314-423-2190
A shareware telecommunications package. (Chapter 3)

SimuComm
Jack Gittinger
University of New Mexico
Center for Technology and Education
School of Education
College of Education
Student Services Center B-69
Albuquerque, NM 87131
505-227-3603
A data retrieval simulator, also through Softswap. (Chapter 3)

The Other Side
Tom Snyder Productions, Inc.
90 Sherman Street
Cambridge, MA 02140
800-342-0236
A game that makes use of telecommunications by having two teams compete for resources and communicate with each other via telecommunications. (Chapter 5)

Window on Telecommunications
Exsym
301 North Harrison Street
Building B, Suite 435
Princeton, NJ 08540
609-737-2312
A tutorial for students to learn the concepts and terms associated with telecommunications. (Chapter 3)

TEACHING MATERIALS

CIP (Classroom Instruction Program)
See *DIALOG* under ***Commercial Networks.*** (Chapter 4)

Ethics On-line!
Exsym
7016 Dellwood Ave.
Albuquerque, NM 87110
505-881-3670
A package designed to aid the teacher in discussing telecommunications ethics. (Chapter 3)

Classroom Integration of Telecommunications: Focus—Bulletin Boards, Electronic Mail, and Databases
Oklahoma State University
Department of Curriculum and Instruction
College of Education
Gundersen Hall 302
Stillwater, Oklahoma 74078-0146
405-744-6279
A videotape of a teleconference designed for training teachers in the use of telecommunications in the classroom. (Chapter 5)

Going Online, a video
Learned Information, Inc.
143 Old Marlton Pike
Medford, NJ 08055
609-654-6266
$15 for 10-day preview, applicable to purchase of video

Roxanne Mendrinos
MENCAP Productions
E. W. Thurston Junior High School
850 High Street
Westwood, MA 02090
617-326-7500
Several units for junior high school, integrating telecommunications, are available from Ms. Mendrinos. (Chapters 4 and 5)

USA/USSR Youth Summit
Massachusetts Educational
Television Lending Library
Bureau of Educational Technologies
The Commonwealth of
Massachusetts, Department of
Education
75 Acton Street
Arlington, MA 02174
617-641-3710
A videotape of a live youth summit between high school students in the U.S. and the Soviet Union, available for a two-week loan at $10.

Glossary

acoustic coupler A device that attaches to a modem to allow the handset of a telephone to connect, through speakers recessed in noise reducing rubber cups, to the modem.

ASCII American Standard Computer Information Interchange—a standard 7-bit binary representation for letters, numbers, and other keyboard and control characters.

asynchronous Data that is organized with start and stop bits included so that either modem can send data to the other whenever it is convenient. See also ***synchronous.***

bandwidth The portion of a communications channel required for effective data exchange. The higher the baud rate, the greater the required bandwidth.

baud rate The speed of data transfer between computers, expressed in bits per second.

BBS See ***bulletin board system.***

binary a base-two representational system usually using the numbers 0 and 1.

bit One binary digit.

Boolean logic See ***Boolean operators.***

Boolean operators The use of the words "and," "or," or "not" to delineate subsets of information.

bridge A link between similar local area networks to manage the exchange of information.

buffer A space in the computer's memory that is used for temporary storage.

bulletin board system (BBS) A telecommunications service for sharing information with the general telecommunications audience.

byte The number of bits, usually 8 or 16, that make up a piece of information.

capture To save electronic data on a disk.
carrier signal The signal a modem sends over a telephone line indicating it has made a connection with a host computer.
CB (Citizen's Band) The term used to describe real-time communications via computers. See also ***chat.***
channel A physical or logical pathway over which information can be transmitted.
chat, chatting Communicating via computers with another person or people in "real time."
chat mode See ***chat.***
circuit The complete path of electric current that performs a specific function within a computer.
circuit board The boards on which electric circuits are attached. See also ***circuit.***
clock card A piece of circuitry placed in the computer to keep track of the date and time, used for bulletin board software. Many computers now have built-in clocks.
command See ***command string.***
command mode The operating state of a modem in which it interprets all data sent to it as instructions that control its internal settings and functions.
command set See ***command string.***
command string A series of key strokes that sends directions to a computer or peripheral device.
computer conferencing See ***teleconference.***
computer network The connection of two or more computers through telephone lines or cables so that people can exchange information.
computer prompt See **prompt.**
control mode The operation of a communication software program when it is using the subroutines that permit the user to manage the setting of modem protocols and computer resources, usually by way of menu options.
data mode The operating state of a modem in which it interprets all data sent to it as information to be exchanged with a host computer.
database A body of related information, usually accessible by a computer.
distance education The use of telecommunications to offer instruction to students in different geographical locations at the same or different times.
distance learning See ***distance education.***
download To store information received from another computer. See also ***upload.***
duplex The capability of computers to both send and receive information at the same time.
echo board A bulletin board system that has the capability to direct files received to other bulletin boards.
echo characters Characters sent to the user's computer screen by either the host computer (full duplex) or the user's computer (half duplex) to show the data being sent.
electronic conference See ***teleconference.***
electronic mail (E-mail) Information sent from one person to another through a computer network.
fiber optics Transmission of information in the form of light waves through thin strands of glass fiber, about the thickness of a human hair. One fiber optic strand can carry many more light wave signals simultaneously than one conventional copper wire using electric current.
file server A device that carries out the transfer of programs, data, and text between devices in a local area network.
gateway A link between dissimilar local area networks.

half duplex The capability of a computer to send and receive information, but not simultaneously.
handshaking The signals that control the flow of data over a modem.
host computer The computer with which you wish to make contact.
information utilities Commercial databases which provide information to their subscribers.
in-line modem A modem that connects directly to the telephone jack, without needing an acoustic coupler for the telephone headset.
Integrated Services Digital Network (ISDN) A data exchange standard being proposed for all local area networks, computer interface ports, and communications software, as well as the telephone system.
interface port The part of the computer designed to allow the computer to communicate with peripheral devices.
LAN See ***local area network.***
load To store a piece of software from a computer disk into the memory of the computer.
local area network (LAN) Computers tied together for communications purposes in a limited geographic area.
log-on, log-off The process of entering and leaving an electronic communications system.
menu A list of options available to the user of a piece of software.
meta-character A combination of keystrokes that simulates keys not actually on the keyboard.
modem Short for MOdulator/DEModulator, a device that translates serial data into tones for transmission over telephone lines and back into digital signal for use by computers.
monitor The computer screen.
network See ***computer network.***
offline Not linked to other computer(s) for telecommunications; generally advisable for preparing material to send over a network and reading information saved from telecommunications, to keep telecommunications charges to a minimum.
online Linked with other computer(s) for telecommunications.
optical fiber See ***fiber optics.***
packet switching A configuration that permits multiple users on any given channel, such as a telephone line, at the same time.
parallel data Data that is organized so that all the bits in a byte of information are handled simultaneously.
parity A means of checking the accuracy of information sent from one computer to another.
peripheral devices Machines or devices that attach to a computer to enable certain functions to be performed, e.g., printers, disk drives, modems.
piracy The illegal copying of software.
port See ***interface port.***
prompt A signal, usually a blinking cursor, letting the user know the computer is waiting for input.
protocol The order of events, such as the settings of the data exchange conditions, for a modem.
save buffer See ***buffer.***
serial data Data that is organized so that one bit is sent at a time.
sofcasting The use of commercial broadcast signals for the distribution of digital data.
start bit A bit added to the beginning of a byte of data to signal the modem that a byte of data follows.
stop bit A bit appended to the end of a byte of data to signal the modem that the complete byte has been sent.
synchronous When the sending of data between two computers is controlled by a single clock, start and stop bits are not needed so data is sent more

quickly, but usually at greater cost. See ***asynchronous.***

systems operator (sysop) The person who manages the software and hardware used in a telecommunications system.

telecommunications Communications among computers across distances by use of radio, television, telephone, telegraph, and computer networks.

teleconference An online "meeting" in which participants interact with each other at the same time or at different times, from different geographical locations, with the aid of a computer network and sometimes satellite sending/receiving stations.

telefinance Financial services augmented by networked computers.

teletype A mechanical machine used in the early years of computers as an input/output device for a computer.

terminal An input/output device without any capabilities of its own attached to a host computer.

terminal emulation Operation of a computer wherein it acts as a terminal.

terminal mode The operation of a communications software program when it is using the subroutines that permit the user to exchange information and data with a host computer through a modem.

text data Characters that have an ASCII representation.

tone pairs The sound or modulation of information as it is sent over the telephone line; relates to baud rate.

transfer protocols Communications software subroutines that allow computers to send and receive data without errors caused by static on the telephone lines.

TXT data See ***text data.***

upload To transmit information for storage by another computer. See also ***download.***

virus A program, often hidden in public domain software, designed to create havoc in any computer into which it is introduced.

Xmodem/Ymodem Methods for checking the accuracy of data transmission, more reliable than the parity check.

Xon/Xoff The standard way, via the keyboard, to start and stop data transmission.

Index

Integrating telecommunications into education.